T0342638

Hailing originally from Canada, Dr Paul E Hardisty has spent thirty years working all over the world as an engineer, hydrologist and environmental scientist. He co-founded international environmental consultancy Komex Environmental Ltd, which he helped develop from a startup to a US$100 million-a-year turnover company with a thousand employees, and sold to ASX-listed Worley in 2007. Paul was CEO of the Australian Institute of Marine Science (AIMS) from 2017 to 2023, and now writes full-time. He is the author of several peer-reviewed journal articles, two textbooks on environmental sustainability, and seven novels. *The Abrupt Physics of Dying* was named *Telegraph* Thriller of the Year and was shortlisted for the CWA John Creasy New Blood Dagger. His latest bestselling novel, *The Forcing*, and its prequel, *The Descent*, imagine a near future affected by climate change. *The Forcing* was shortlisted for the 2023 Crime Lovers' Awards. Paul lives in Western Australia.

DR PAUL E HARDISTY
IN HOT WATER

affirm
press

affirm press

First published by Affirm Press in 2024
Bunurong/Boon Wurrung Country
28 Thistlethwaite Street
South Melbourne VIC 3205
affirmpress.com.au

10 9 8 7 6 5 4 3 2 1

Affirm Press is located on the unceded land of the Bunurong/Boon Wurrung peoples of the Kulin Nation. Affirm Press pays respect to their Elders past and present.

Text copyright © Paul E Hardisty, 2024

All rights reserved. No part of this publication may be reproduced without prior written permission from the publisher.

 A catalogue record for this book is available from the National Library of Australia

ISBN: 9781923022386 (paperback)

Cover design by Luke Causby/Blue Cork © Affirm Press
Author photo by Donna-Lisa Healy
Typeset in Garamond Premier Pro by J&M Typesetting
Proudly printed and bound in Australia by the Opus Group

For my sons, Zachary and Declan

Foreword

After putting the Australian Institute of Marine Science back on its feet, Paul Hardisty has served his term as CEO and is now free. To do what, we all wondered? We should have guessed, for this book had obviously been brewing inside him for a very long time. It must have been hard to keep the lid on, given that the Great Barrier Reef crisis was in his face every day from the beginning: the science, the politics, the people. Hundreds of people with endless opinions, demands, wants and needs, professional and personal. And so this book, as with its many predecessors, is steeped in history and, as it moves on, doom and gloom. But its theme, perhaps surprisingly, is one of hope – not a wellwisher's hope, for that leads nowhere, but hope that springs from deep enquiry, deep thinking and, ultimately, clear understanding.

As a fresh face in reef science, Paul brought new ideas and new hope. He has given many reef scientists the means and motivation to redouble our efforts to bring the Great Barrier Reef through the troubled years that lie ahead. Some say that this is a waste of time and money as the reef is a lost cause. They might be right; Paul explains his personal view of this. He accepts that international efforts to contain global temperature increase, the root cause of the bleaching that is killing corals, are unlikely to stave off critically damaging conditions for coral reefs. But even so, he finds hope that we humans can repair the damage we have caused. My science-centric perspective is that this raises an extraordinary paradox. Corals have been on our planet for 250 million years and in that time have survived two of the greatest extinction events ever recorded. They have survived higher temperatures, far greater greenhouse gas levels and more acidified oceans than our worst predictions. They have adapted to one of the most hostile environments on earth – places where land, sea and air all intersect – to build reefs teeming with life where no other ecosystems have ever survived. And yet coral reefs, the great survivors, are failing in the face of climate change ahead of any other major ecosystem in the sea or on land. We know the key ingredient is time, for we are changing our planet's climate faster than the genetic memory of any coral. That's where Paul's human perspective meets mine. We both believe that if we could buy corals enough time to adapt, they would surely make it. We know that there are people on all fronts who understand what needs to be done. We meet some of these people, the good and not so good, in the pages of this book.

Please allow me a moment's reflection. The Great Barrier Reef, nature's pinnacle of achievement in the ocean realm, is the embodiment of wilderness, of remoteness – a place of endless beauty that has endured when so many other World Heritage places on Earth, cherished by generations past, no longer engender strong emotions since they have been altered beyond recognition. Most Australians have never seen the Great Barrier Reef, but all know it is our icon, a special place that is quintessentially Australian, a place that we are proud of and boast about. Will we still have this view twenty years from now? Or will we be having conversations wrapped in regrets and starting with 'if only'. Twenty years ago, worried scientists like me could be dismissed as being overly pessimistic, and dismissed we often were, especially if our utterances were seen to be harmful to tourism. But we were not pessimistic, or optimistic

– we were simply accurate. Scientists love to be accurate, but it is no consolation that we can now dispel any notion that we once exaggerated, for facts have now replaced prophecies and the fact is that the destruction of the Great Barrier Reef can reasonably be described as imminent.

We now have our last chance and that is what motivated Paul to write this book. It reaches to the hearts of Australians with a simple message: it's now or never.

J. Veron

Charlie Veron OAM,
Former Chief Scientist,
Australian Institute of Marine Science

Contents

List of Acronyms

ACF	Australian Conservation Foundation
AEF	Australian Environment Foundation
AIMS	Australian Institute of Marine Science
ANAO	Australian National Audit Office
ASX	Australian Securities Exchange
CSIRO	Commonwealth Scientific and Industrial Research Organisation
GBR	Great Barrier Reef
GBRC	Great Barrier Reef Committee
GBRF	Great Barrier Reef Foundation
GBRMPA	Great Barrier Reef Marine Park Authority
GDP	Gross Domestic Product
ICRS	International Coral Reef Society
IPA	Institute of Public Affairs
IPCC	Intergovernmental Panel on Climate Change
IUCN	International Union for the Conservation of Nature
JCU	James Cook University
LNG	Liquefied Natural Gas
NOAA	United States National Oceanographic and Atmospheric Administration
RRAP	Reef Restoration and Adaptation Program
UNESCO	United Nations Educational, Scientific and Cultural Organization
WWF	World Wide Fund for Nature

Introduction

If you have ever seen the Great Barrier Reef, you will know that you have witnessed something that transcends beauty. Recognised as a World Heritage Area by the United Nations, it is one of nature's most spectacular treasures – a sparkling necklace of blue sapphires strung along a glittering shore.

The GBR is the largest and most biodiverse reef complex on the planet. Stretching along 2300 kilometres of Queensland's coast, it covers an area the size of Germany. Hundreds of individual reefs and islands are home to thousands of species of coral, fish and birds. Every year it attracts millions of visitors and generates billions in economic activity, supporting hundreds of thousands of jobs.

Everyone wants a piece of the Great Barrier Reef. Ever since European settlement, we have coveted its beauty and riches. We exploited its abundance, harvesting shells and turtles and fish and pearl oysters, sometimes to near exhaustion. As our affluence grew and travel became easier, more of us wanted to experience the reef for ourselves. We built roads and towns and resorts and came in such numbers that we began to destroy the very thing we were there to see. We saw the potential beneath the reef, too, and sought to drill for oil and gas. And as the pressures grew, a small number of visionaries and scientists began to realise that the reef was finite, and fragile. They warned that in our ignorance we were risking doing the reef irretrievable harm.

These warnings, and the efforts of dedicated conservation volunteers, journalists, members of the public and enlightened politicians, sparked a popular movement in Australia and around the world to protect the reef, culminating in 1975 with the creation of the Great Barrier Reef Marine Park. Since then, the pressures on the reef have only increased. And now, a quarter of the way into the 21st century, a new and deadly threat to the reef has entrenched itself: climate change.

~

I grew up near the sea, on the west coast of Canada. I loved swimming in the cold Pacific, sailing the Gulf Islands and exploring the coast. I'd read about the Great Barrier Reef, seen photos, but I never imagined that one day I would be living and working there. Early on, I decided to devote my life to protecting the natural world, and to writing about it. I studied science and engineering. While doing my master's thesis in Ghana, I began to realise that if you want to protect the natural world, you have to first look after people.

After completing a PhD in environmental engineering at Imperial College London, I went back to Canada and co-founded an environmental consulting company. We started helping governments, oil companies and mines manage their waste and clean up their spills. Protecting water supplies, rivers and wetlands became our speciality. We branched into marine science, conducting environmental assessments and helping decommission offshore platforms. We grew as environmental awareness and regulation grew through the 1980s and 1990s, and we expanded into the United States, Europe and South America.

I moved with my young family to the United Kingdom and built our operations there, and then we moved again to Cyprus, from where I ran our burgeoning European, Middle Eastern and African portfolio of work. I was in Ethiopia as the Mengistu regime fell, saw the rich flee as the rebels approached, saw the poor trampled and made destitute. I watched as Yemen descended into civil war, and tried to warn the country's leaders against the rapid depletion of groundwater aquifers in this driest of lands. Everywhere I went, the direct connection between the plight of people and the condition of the environment was apparent. I came to see corruption and misinformation as a scourge on both, sapping wealth and leaving a trail of environmental destruction in its wake. And inevitably, it was the poorest among us, the voiceless, who suffered most.

In 2017, after three years running the climate adaptation science program for CSIRO, and then its Land and Water Business Unit, I joined the Australian Institute of Marine Science as its CEO. I moved to northern Queensland with my wife. Quickly I came to realise that while Australia is by any measure a rich country, the connection between people and the environment applies as much here as in the poorest country in Africa or Asia. In the 21st century, the Great Barrier Reef exists at a tense economic, social, cultural and environmental intersection. Queensland is one of the biggest fossil fuel exporters in the world, and much of the coal and gas it produces is shipped from ports on the GBR. Agriculture employs thousands of people in coastal Queensland, and rainforests adjacent to the reef continue to be cleared for pasture and sugar cane, increasing the flow of nutrients and sediment to the reef. Tourism and recreation are among the state's biggest employers, and the GBR itself generates over A$6 billion a year in economic activity. The Traditional Owners of the land and sea country connected to the reef continue to struggle for a voice in the conversation. Each of these groups wants a say in how the reef is run, accessed, managed and protected.

The closer I got to the issues, the more interested I became in the reef's recent history. What challenges have arisen in the past, and how have they been met? How did we come to create one of the largest marine parks in the world? And why, after everything that has been accomplished, are we now on the edge of a precipice, looking down into the abyss? For despite what people may hear in the popular press, the reef is in deep trouble. And while all of the other pressures on the reef continue to mount, it is the inexorable build-up of heat in the world's oceans which is now the major existential threat. The reef is, quite literally, in hot water.

This book is about the meaning of the reef, its importance to Australians and the world, what has been done to understand and protect it, and what needs to be done now if we want our reefs to survive this century. For if they can make it through to the year 2100 in reasonable condition, there is every probability that they will be with us for a very long time to come. This book is also a warning about what will happen to the reef if we continue on our present course. Faced with the biggest threat to the reef ever, and the most difficult one to address, how will we respond?

The book is structured as a dual narrative. One offers a personal, inside view of the struggles between the key actors attempting to shape the future of the reef since the major back-to-back bleaching events of 2016 and 2017 that ravaged the GBR. As CEO of AIMS during that period, I was close to the action and privy to all of the latest science, unalloyed and unfiltered by the media and social commentators. The second narrative is an exploration of the history of the reef since the start of the 20th century, when scientific discovery on the reef began in earnest, and the urgent need for conservation and protection began to emerge. At its core, this is a story of conflict. Of battles between those who seek to exploit the reef, and those who want to protect it. Of proxy warfare between those determined to slow or even halt progress towards a low-emissions future, and those who understand that climate change threatens everything humans aspire to.

For it is no exaggeration to say that without a healthy environment, life on earth for human beings will become increasingly difficult, and eventually impossible. The fate of the Great Barrier Reef is now inextricably linked to the fate of the planet, as ours is to both. The reef is a symbol of our struggle to rise above narrow self-interest and act for the common good. If we can safeguard the reef, we can save the rest.

I invite you, dear reader, to decide for yourself the goals we should set, the course we should take, and the likelihood of our success.

1

Into the Fray

It was November 2016, and in the aftermath of what would prove to be the first of two back-to-back mass bleaching events that together killed off as much as half of the coral on Australia's famed Great Barrier Reef, a well-known politician floated facedown in the azure waters just off The Keppels, near Rockhampton in Central Queensland. After a moment she raised her head and held up a healthy brown *Acropora* coral for the cameras. The uprooted colony hung dripping and forlorn in her hand, as if surprised to be singled out as an exhibit in this most recent skirmish in the Hundred Years' War for the reef.

Senator for Queensland Pauline Hanson pushed her mask up onto her forehead. 'If you actually go deeper,' she said, breathless, treading water, 'twelve to twenty metres deeper in the ocean, the reefs there are in pristine condition. And they're growing all the time.'[1]

Other members of her One Nation party nodded as she spoke, staring out through fogged masks. They had come, they said, to support the local tourism industry that relied on the reef. Tourism here was way down, they said, mainly because of widespread and false media reports that the reef had been trashed by climate change.

'Which is complete nonsense,' said Malcolm Roberts, another One

Nation senator, elected to Australia's upper house with seventy-seven primary votes (no typo there), the lowest number ever for any member of Australia's parliament. 'All I need is to see the empirical evidence. And there is none.' His glacier-blue eyes flashed with contrarian zeal. According to his party, climate change, which most reef scientists say is driving the marine heatwaves which caused the bleaching and coral death seen across large swathes of the Great Barrier Reef the previous summer, particularly in the north, is nothing more than an elaborate hoax.[2]

Back on the tourist boat after the snorkel, the party's leader fronted up to the TV cameras, her carefully dyed atomic tangerine hair plumped by the sea air. 'The main political parties have been dominated and controlled by the Greens for too many years,' she said, referring to the country's ideological counterweight to her own fringe party. The Greens held a grand total of one out of the 150 seats in the House of Representatives, and nine out of the seventy-six Senate seats. Exactly how they had been able to control the major parties she did not say. Her own party held no seats in the House, and four in the Senate.

One of the journalists in the party asked Ms Hanson about the reports of mass coral death on the reef.

These reports had been verified by some of the top coral reef scientists in the nation, including Professor Terry Hughes, director of the prestigious Centre of Excellence for Coral Reef Studies at James Cook University. Hughes conducted the airborne surveys himself, flying the length and breadth of the reef over several weeks in the aftermath of the heatwave, carefully documenting the bleaching. Hughes knows coral – he has been studying them for most of his adult life. His PhD from Johns Hopkins University was on the resilience of Caribbean reefs. After postdoctoral research at the University of California, he moved to JCU in Queensland in 2000, and is now recognised as one of the world's most esteemed coral reef scientists. His resume is as long as your front drive and includes hundreds of publications in the most respected scientific journals.

The results of Hughes' airborne surveys had been picked up and amplified by news outlets across Australia and around the world. The headlines varied, but the message was stark: the Great Barrier Reef, a UNESCO World Heritage–listed place of outstanding universal value for all of humanity, was under serious threat. And climate change was the number-one culprit.

Senator Hanson's answer to the journalist's question was immediate and pointed. 'We can't have these lies put across by people with their own agendas,' she said. 'We are being controlled by the UN,' she continued, pressing home her point. 'And these agreements that have been done, uh, for people's self-interest ...' She faltered here a moment, seeming to have lost track of what she was saying, but found the thread again. '... And where they are driving our nation as a sovereignty and the economics of the whole lot.'

The journalists held there a moment, blinking in the tropical light that shimmered through the ship's cabin, looking confused and unsure how to proceed. After a while it became clear that the press conference was over. No one bothered to ask the senator about her party's positions on climate change or land clearing, two of the issues scientists say are contributing to the problem. Nor did they point out that the piece of reef they had just visited was over a thousand kilometres south of where the main bleaching has occurred. There was no need. Everyone already knew. And nobody mentioned that ripping live coral from the sea floor in a marine park without a permit is prohibited.

The boat headed back to shore, where the guests disembarked. The tour company owner seemed satisfied that his concerns had been heard. A few hours later the first media reports hit the online news sites. The reactions were as scathing as they were predictable.

A leading global NGO dedicated to the preservation of nature called for the senator to actually visit a part of the reef that *had* bleached and see for herself the impact that warming ocean temperatures were having. A member of the Climate Council suggested that 'the trip is like taking journalists reporting on a conflict to a five-star holiday resort miles away from the actual war zone'. Another NGO suggested that she was in denial about the health of the GBR, and that she should 'disembark from the SS *Ignorance* and come back to shore'.

In a few days, this exchange would be replaced by some new, more violent, and equally fruitless confrontation. For more than a century, the fortunes of the reef, its creatures and those who have a stake in its future have ebbed and flowed. But here in late 2016, stalemated sides were dug in along increasingly deep and reinforced lines, engaging in a slow battle of attrition in which the casualties mounted up, not only in the form of truth obscured, but in the forlorn shapes of public confusion, lost opportunity, wasted money, vanishing species, and vast fields of dead and dying coral. As I would soon learn, this was a war that had been going on a long time.

~

The GBR had experienced widespread bleaching (sometimes termed 'mass bleaching') before, for the first time in 1998, again in 2002, and most recently in 2006 when reefs in the south of the GBR, and particularly around The Keppels, were hit. But the 2016 event was different. Not only was this the most extensive bleaching the reef had ever experienced, it was also the most devastating. The media coverage was intense. Suddenly, the plight of the Great Barrier Reef, and the fact that climate change was the culprit, was getting global attention.[3] And with the scrutiny came the politics. It was German historian Carl von Clausewitz who famously described war as 'a continuation of politics by other means'. In 2016, the Great Barrier Reef was as political as it gets, and the weapon of choice in this war was disinformation, with the favoured means of delivery the media.

At this stage, I was watching from the sidelines. I'd just resigned from CSIRO after three years first running the Climate Adaptation Flagship, and then, when that was wound up following an organisation-wide restructuring, the Land and Water Business Unit. I was working on a novel and spending some time travelling overseas. But wherever you happened to be, coverage of the bleaching event was inescapable. All the Australian dailies ran stories, each predictably slanted to their own political bias. Social media commentators confused everyone, with some claiming the GBR was already dead, and others that it was as healthy as it had ever been. Each side rolled out its own grey-haired experts and eminent scientists to support its claims. For the average person who was not a scientist and who may never even have visited a coral reef, it was hard to know what to believe.

The coverage quickly went international. Days after the senators' snorkelling trip, the BBC's headline 'Great Barrier Reef suffered worst bleaching on record in 2016, report finds', was backed up by extended TV reports showing underwater footage of bleached and dying corals. A day later, the New York Times led with 'Great Barrier Reef Hit by Worst Coral Die-off on Record, Scientists Say'. The stories described the airborne surveys by JCU, and quoted reef scientists like Hughes and University of Queensland Professor Ove Hoegh-Guldberg, one of the first scientists to predict that bleaching would become more frequent and so pose a clear threat to the future of reefs around the world.

But no one quite hit the crescendo of doom reached by *Outside* magazine a few weeks before the senatorial reef visit. The popular magazine for outdoor enthusiasts – exactly the kind of people the tourism industry hopes to attract – published a mock obituary for the GBR. These were its opening lines:

> The Great Barrier Reef of Australia passed away in ... after a long illness. It was 25 million years old.
>
> For most of its life, the reef was the world's largest living structure, and the only one visible from space ... Among its many other achievements, the reef was home to one of the world's largest populations of dugong and the largest breeding ground of green turtles.

It concluded:

> The Great Barrier Reef was predeceased by the South Pacific's Coral Triangle, the Florida Reef off the Florida Keys, and most other coral reefs on earth. It is survived by the remnants of the Belize Barrier Reef and some deepwater corals.
>
> In lieu of flowers, donations can be made to Ocean Ark Alliance.[4]

Although clearly intended as a warning, you could see why some tourism operators might be upset.

At home, *The Australian* ran a carefully worded piece with the subheading 'The bleaching of parts of the reef is dividing the scientific world'. The web version was headlined 'Great Barrier battleground over coral bleaching'. The message was clear: scientists were divided over what was happening. The battle lines were drawn.

It was about then that I was approached by headhunters tasked with looking for the next CEO for the Australian Institute of Marine Science (AIMS). My first reaction was to ask if they had the right person. I am an engineer, not a marine scientist, and although I had worked in related areas, I had certainly never considered a role leading one of the premier marine research organisations on the planet. The recruiter said that they were looking for someone who could lead the strategic rejuvenation of the institute and build its long-term financial sustainability, and they considered my background and experience ideal. I was persuaded to have a look, and the more I did, the more intrigued I became. AIMS had tended to keep a relatively low profile, but had managed to build a fifty-year track record of world-leading marine science, was highly respected and blessed with a state-of-the art campus and facilities near Townsville in North Queensland, as well as two other laboratories in Darwin and Perth. I applied and not long after, to my surprise, I was told I had been successful.

~

The causeway out to the AIMS headquarters at Cape Ferguson ran dead straight through shimmering tidal flats and ribbons of fringing mangrove towards the rocky hills of the peninsula. Giant cranes – the native brolgas – waded in the shallows. Pelicans wheeled in a cloudless sky, a half-dozen of them, graceful and detached. I was thinking about what awaited me, about everything that had happened on the reef since I had agreed to take on the role as AIMS CEO.

In describing what had been going on, I don't use the word *war* lightly. Words matter, and I have seen my share of conflict: real war, the kind where politics has failed, men take up arms and people die. Sun Tzu, the 5th-century Chinese philosopher, said that all warfare is based on deception. But even in this war of words, truth exists. For words can both bury and exhume the crimes of thought and omission, the guilt of soft inaction, and the venal shame of self-betrayal.

As an immigrant to Australia, I was still coming to terms with my adopted country. Life is good here, in the so-called lucky country, if you have some money and like the outdoors. It is a place of opportunity, extremes and deep contrasts. The land and seascapes are beautiful and often unforgiving, the spirit of the people open and generous. But there is another dimension, too – harder to get at, hidden beneath the surface. A brutal past. Deep social divisions. Plus, as I was finding, a cynical and divisive political environment, and a disturbing lack of concern by some for the condition of the country's unique plants, animals and wild places.

The country's emotional relationship with the ocean is undeniable. Australia's marine estate is the third largest in the world, covering over thirteen million square kilometres – almost twice the area of our continent. The majority of our population lives within a hundred kilometres of the coast, and the oceans are a big part of our national culture and identity. For the coastal Indigenous communities, their connection to sea country goes back tens of thousands of years. The value of our blue economy – the total value of goods and services produced from, in and for the oceans – exceeded A$118 billion in 2021, contributing over 4.5 per cent of the nation's GDP, and employing almost half a million people in full-time jobs.[5] That's more than Australia's coal sector, and more than its agricultural sector.

After a decade here, raising two sons with my wife, Heidi, I'd seen a lot of the country, met all kinds of people. And yet, Australia still found a way to surprise me almost every day, and to remind me of how little I knew. And now in a new job, I was about to be surprised again – but this time in a way that would fundamentally change the way I looked at the world.

In the car on my commute for my first day at AIMS, the open flat country gave way to the thick forest of Bowling Green Bay National Park, and then the first pillared granite outcrops of the hills. Forty-five minutes out of Townsville, I reached the front gate of the Australian Institute of Marine Science, the nation's national marine research agency. It was hot outside but not unpleasant. The oppressive humidity of summer was still a few months away. The barrier opened. I took a deep breath, pushed down the blue swarm of iridescent butterflies whirling in my chest, and drove up the hill.

The road here had been blasted through the hard granite that formed this peninsula hundreds of millions of years ago. You can still see the drillholes in the rock where the dynamite was set. The undergrowth in the forest on either side of the road was yellow and dry. It hadn't rained properly here for almost two years and Townsville's dam was nearly empty – I'd heard on the news that the city was running out of water. As I reached the crest of the hill, I slowed the car to a stop. There before me, spread out as far as I could see, were the blue Pacific waters of the Great Barrier Reef.

It was July 2017, more than six months since the senators' boat trip to The Keppels and the acrimonious Senate debate, and four months since then-treasurer Scott Morrison famously held up a chunk of anthracite in parliament during an attack on Labor's energy policies, declaring, 'Mr Speaker, this is coal. Don't be afraid. It won't hurt you.'[6] But the problem was, unfortunately, that coal *was* hurting us. The GBR had just suffered its second mass bleaching event in as many years, doubling down on the destruction of 2016, with climate change as the major culprit – driven in part by the burning of coal. And since then, the battle cries from both sides had only intensified. I swallowed my fear and drove down towards the main buildings.

14

As a newly appointed leader, from the first moment you walk in the front door, you are on trial. Everyone will be watching you, weighing up everything you say. Every member of staff will be judging you, wondering what you will do, how you will change the place and what it will mean for them. It's a daunting feeling. I closed the car door, straightened the collar of my jacket, and walked across the car park towards the main doors, heart galloping. I was not ready for this, not even close.

At the time, my first-hand knowledge of Australia's largest coral reef system was limited to a holiday visit in 2007, the year after we immigrated to Australia as a family. We had sailed around the Whitsunday Islands for a week. The weather had been good, and the water clear. We had snorkelled from the boat, marvelled at the rivers of colourful fish flowing through deep forests of branching coral. I had done some background reading, of course – several papers, some articles on the web – but it had all happened quickly. Of what had caused the coral bleaching that I had heard so much about in the last few months, I knew only the basics. The deeper secrets of a reef's life were a mystery to me; I couldn't name a single species of coral other than by its common name – brain coral, staghorn.

But what I lacked in coral expertise, I made up for with a lifelong passion for the natural world, and especially the sea. My birthday present in 1972 at age ten was a book by French oceanographer and explorer Jacques-Yves Cousteau, *Life and Death in a Coral Sea*. Full of amazing full-colour photographs taken underwater, the book fired my imagination like nothing quite had before. The book not only opened my eyes to the wonder and beauty of the hidden world beneath the surface of the ocean, but to the dangers that pollution and over-exploitation posed. Sitting in my upstairs bedroom in our house in Toronto, watching the snowflakes fall past my window, I wanted to be a marine scientist. My parents knew exactly what they were doing, because that same Christmas they took my brothers and me to the Caribbean for the first time, to a small coral cay in the Bahamas.

My first experience snorkelling a reef was there, on Chub Cay in 1972, at the age of ten. The coral was right there, and so were the fish. Wade a few steps out from the beach, stick your face in the water, and there it all was. Silver barracuda slipped silently between huge coral domes patterned like diagrams of the human brain I'd seen in books. Giant conch shells, the curl of their inner edges glistening pink enamel, littered the sandy patches between shocks of branching coral. Fish of every colour and pattern imaginable – and quite a few, such as the eerily changeable porcupine fish, beyond my imagination – went about their business in a city whose architecture was defined not by the straight lines and ordered rows of doors and walls and roofs of my world, but by an infinite complexity of entirely foreign branches and plates and fans and domes. I was Cousteau, exploring a beautiful and untouched world. It was pure magic. I watched a stingray give birth to live young in the shallows of that little island, and watched the baby swim away under the protective wing of its mother. I fell in love with nature and the sea.

But by 2017 coral reefs in Caribbean were in bad shape.[7] A combination of coral disease, devastating hurricanes, marine pests and a series of now worryingly familiar bleaching events due to elevated sea temperatures was causing severe declines in coral cover from the US Virgin Islands to the Antilles. I was heartstruck at the thought of what had become of that magical childhood place I had explored as a wide-eyed boy almost half a century before. It didn't seem that long ago. I was determined to do everything I could to help our coral reefs survive and flourish in an increasingly hostile world.

2

Why Care?

On my first day at AIMS, I met with a group of scientists for an introductory briefing. They were from all over the world, women and men, some with decades of experience riven in their sun-worn faces, others mid-career, young still, but highly trained and confident. We crowded into the main conference room. The lights were dimmed, and the briefing began.

Over the past two summers, they explained, the reef had lost as much as half of its shallow hard coral cover. They showed me diagrams of the extent of the bleaching compiled from consecutive aerial surveys. The full scale of the damage wouldn't be known until they could get in the water and see it with their own eyes. There was only so much you could tell from the air. Later in the year, when the long-term monitoring program was completed, the datasets on coral cover would be updated. They spoke in measured tones in their scientists' language of facts and data, hypotheses and processes. These people were trained to be rational and objective. Their credibility depended on it. But there was an edge in their voices that betrayed an alarm that didn't come through in their precise, clinical sentences.

'We have been monitoring these reefs for decades,' said Dr Britta

Schaffelke, the most senior of the scientists. Originally from Germany, her accent had softened after almost two decades in Australia. She ran the AIMS research program that focused on the GBR. Her fair greying hair framed a pair of overcast eyes set in a lean, symmetrical face. Worry stretched her sun-browned forehead and held her lips tight. 'Our records go back to 1985. And we've never seen anything like this before. Not even in '98.'

That year, 1998, is one everyone here remembered or had heard about. An intense El Niño, the natural ocean cycle which brings warming conditions to the Pacific, triggered the first recorded mass bleaching on the GBR. It also hit coral reefs worldwide. The reef succumbed to bleaching again in 2002, albeit less severely. In between, there were cyclones, which can literally rip coral from the sea floor, and there were outbreaks of the coral-eating crown-of-thorns starfish. They promised me more detailed briefings on those things later, but right now, the focus was on the clear and immediate threat.

'These ENSO cycles are natural,' Britta continued, assuming that I knew the acronym stood for El-Niño–Southern Oscillation. I didn't. I stopped her, asked her to explain. Her tone was patient and precise as she described how climate change was heating up the oceans, which meant these natural oscillations were now superimposed on a baseline which was shifting up, pushing the peaks to record levels. That was what we were experiencing now. We no longer needed an El Niño to get bleaching. And we'd never had mass bleaching in consecutive years before.

After the briefing I was taken on a tour of the National Sea Simulator, or SeaSim, the world's largest and most sophisticated research aquarium complex. Nestled among the rocks of the headland, this futuristic facility was the brainchild of David Mead, the Institute's chief operating officer. An engineer by training, David imagined a place where scientists could run experiments of long duration and extreme precision. 'We saw how scientists had been doing their experiments – literally going to Bunnings, buying some bins, pipes, and pumps, and setting it all up themselves. A cottage industry. We knew from the large-scale industrial process industry that there was a better way.'

It had taken Mead and his colleagues years to convince government, but finally funding was secured through a program called the Super Science Marine and Climate Initiative, and construction was begun. The facility was opened in 2015 by Senator Kim Carr, then science minister and a key backer of the project. Inside, the facility hummed with quiet efficiency. Small experiment rooms sparkled with banks of small glass aquaria, connected by multi-lane highways of coloured pipes and cables, each one carrying treated and prepared seawater controlled to exact temperatures, and dosed with precise concentrations of key compounds. 'The big difference here is that we can control everything to extremely fine tolerances,' said Mead. 'Out there on the reef everything changes second by second. During the day the water warms, corals photosynthesise, producing oxygen. As night falls photosynthesis shuts down and the corals give off CO_2. Temperature, pH, partial pressure of CO_2, all of these things are changing continually. If you aren't capturing those cycles, you're not reproducing the actual conditions on the reef. And if you're not doing that, your experiment isn't going to be as good as it needs to be.'

We continued through the building and came out onto an elevated platform. Spread below us was a huge room filled with large tanks, each connected to a maze of pipework and lit above with banks of carefully calibrated lights. Mead explained that lighting was key. 'We need to reproduce the same spectrum as out on the reef, and vary it to match. If you come back here at night, it's dark here. The corals have gone to bed,' he said, grinning.

We headed down the stairs and stood in front of a huge display tank. Fish of every colour swam within a forest of coral of every form and hue. Near the front, a family of orange-and-black clownfish – Nemo from the children's movie – patrolled the swaying arms of their anemone home. The colours were so vivid it all looked fake, like an overblown computer-generated image from a movie. 'We put this in so people can see what a healthy outer reef on the GBR looks like,' Mead said.

Next to it was a smaller tank that looked very different. Dead coral lay in piles of grey rubble, covered over with thick green turf algae. A few fish circulated as if looking for something they had lost. 'That one shows people what a dead reef looks like,' said Mead.

Beyond the display area we moved through the main experimental areas. Dozens of huge tanks were set row on row, each filled with different types of coral at various stages of development. We came to a series of tanks with the words 'Assisted Evolution 2015' written on the glass in wax pencil. This experiment was one of the reasons they needed to build the SeaSim, Mead explained. They'd known since 1998 that they were going to have to look into whether they could selectively breed corals to be more heat tolerant. These were Professor Madeleine van Oppen's experiments, he told me. She had been studying the genetics of corals and their symbionts for years, and her eerily prescient work had already attracted international attention.

'One day, if things keep going like they are, we might have to think about putting heat-adapted corals back out onto the reef to help it cope,' said Mead. His team was already developing a plan to fund the research needed to make this happen, should it be required.

We finished up the tour and shook hands. Back in the main building, I met one of the younger scientists, Dr Neal Cantin, a marine biologist who'd moved to Australia to study coral reef biology at James Cook

University here in Townsville, and joined AIMS after completing his PhD. A fellow Canadian, it turned out he was a fan of the Toronto Maple Leafs ice hockey team, like me. We exchanged the usual wry commiserations: our team hadn't won a Stanley Cup, ice hockey's oldest and most famous trophy, since 1967. At the time of publication of this book, Leafs fans' misery continues.

Back to business, Cantin showed me the Institute's coral core library. He explained how these cores are collected, by drilling into the massive dome-like *Porites* coral to retrieve their cores, which can be dated much like the ice cores climate scientists cut in ancient glaciers. These corals grow slowly, he explained, a few millimetres a year only. But they are tough and long-lived, perfect for this kind of analysis.

His delivery was clear, without emotion, almost laconic. This was his work, explaining the basic elements of what he and his team do, how they do it and what it means. He switched on a UV light and pointed out the banding, each layer representing a year of growth. It reminded me of the tree rings in the stumps of the big pines felled in the bush near where I grew up in Canada.

Some of the cores stored here went back over four hundred years. Cantin showed me one such specimen, on display in the building's main hub. Time markers had been set out against the dusty white of the core which stretched from the floor all the way to the vaulted ceiling some six metres above. The marker nearest the floor read: *1522 – First circumnavigation of Earth by Ferdinand Magellan*. A metre or so above was another marker: *1770 – James Cook charted east coast of* Terra Australia Incognita, *named Cape Cleveland*. I raised my gaze again, and the subtle bands in the coral accreted the years: *1846 – James Murrells (Morrill) shipwrecked on the beach just outside here, spent 17 years living with Aboriginal people.* You had to raise your chin to look up now, the reef building inexorably silent in its crystal-clear sea, year after year: *1936 – Eddie Mabo, Indigenous rights champion, born in Torres Strait*. And then, right at the top, near the ceiling: *1985 – this core collected from Sanctuary Reef, Southern GBR*.

Cantin brought my attention to a smaller sample nearby. He traced his finger along the core's length until it came to rest on a dark jagged band. It was almost as if the core had been broken here, and then glued back together and scarred over.

'This is 1998, the El Niño,' he said, explaining how this part of the colony bleached and died. It then took a few years for the surviving parts of the colony to grow back and heal over the dead tissue. There is nothing like this before 1998 in any of the over ten thousand cores AIMS has collected from reefs around the world.

'So this is real,' I said, thinking of the two contrasting display tanks in the SeaSim. 'The bleaching, I mean.'

He looked at me for a moment as if I had come to the wrong place. Maybe I had.

'Yes,' he said. 'It's happening.'

I asked him what he thought the future would hold. He looked down at his feet. This was not comfortable territory. I watched him take a deep breath. The scientific community had been trying to warn the world about this for years, he explained. If we didn't change what we were doing, and change it fast, we might not have any reefs left by mid-century. He stopped, stared at me as if to say, *Surely, surely, that is enough*.

~

Within days of starting at AIMS I was being interviewed on radio. What was happening out on the reef? Was it really dying? Why should people care? 'Keep it simple for the listeners,' I was told just before going on. It was a live broadcast. The speaking notes provided to me by Britta Schaffelke and her team trembled in my hands. I was learning as I went.

I started by explaining AIMS's long-term monitoring of the reef and the trends we were seeing, and moved quickly into the back-to-back bleaching events. Both the 2016 and 2017 bleachings had been caused by marine heatwaves that hit the northern third and then the middle third of the GBR respectively. Corals are extremely sensitive to temperature. They have existed for thousands of years precisely because they have enjoyed long periods of relatively stable water temperatures. In fact, if ambient water temperature increases by more than a single degree over the normal upper limit of their comfortable range, and is maintained for more than four weeks, many tropical coral species start to reject their algal symbiont – the zooxanthellae that provide the coral with most of their energy and their wonderful colours. If water cools at that early stage, because of increased cloud cover, cooling rains or the appearance of a cyclone, the symbiont can return and the coral will recover and survive. But if higher-than-normal temperatures persist for eight or more weeks, the coral will start to die.

'Is the reef really dying?' the interviewer asked.

I explained that parts of it had died, yes, but overall, at this point in time, the reef was alive. 'The reef is huge,' I said. 'It's battered and bruised, but it's still beautiful, and it's fighting hard. And it needs our help.'

'So it can recover?'

'Yes,' I read from my notes. I talked about our work on isolated reefs in Western Australia that showed how a badly degraded reef can recover in about a decade. But the recovery was not a straight line. For the first few years after a major die-off you didn't see much change. Then slowly, over the years, the recovery built, until by years seven, eight and nine there was an explosion of growth. As long as there is enough resilience left in the system overall, reefs can and do recover.

The interviewer asked me about other pressures on the reef. What about water quality? Over the last several years both the Queensland and Commonwealth governments had put huge amounts of money into improving water quality, through programs designed to help farmers and graziers reduce the flow of sediments, fertilisers and other chemicals onto the reef. I had been given a copy of the 2017 Consensus Statement on Water Quality, a summary of what was currently known about the effects of declining water quality on the reef. I was only part way through it, but the report concluded that farming, mining and urban development throughout the GBR catchment area had, over the years, caused a decline in water quality on the reef. Agricultural development had altered river flows, sediment and nutrient loads. Most catchments and floodplains up and down the coast had been extensively modified, including by the construction of dams, large scale clearing of native vegetation, disturbance by livestock, and extensive use of fertilisers and pesticides.[1] The report had been prepared by some of the best scientists in Australia, representing many different organisations, and had been endorsed by the Australian Academy of Science.

It was already clear to me from my briefings and reading that while we needed to do all we could to improve the health of the reef – including reducing the impact of poor water quality – it was the effects of rising ocean temperatures that had now emerged as the single biggest threat to the reef. The northern part of the reef, historically the least affected by water quality issues, was the worst-hit by the most recent bleaching. A few weeks of extreme heat had undone all of the years of intensive effort to improve water quality.

'What if the bleaching events come more frequently than once in a decade?' she asked me. 'Like just now, in successive years?'

I said something like: 'This is a concern. We know that the ocean is taking up the majority of the accumulated heat in the earth's system right now. If bleaching comes before the corals get a chance to recover, well, there are only so many hits anyone can take before they don't get up again. These bleachings are a warning sign of what is to come if we don't get serious about reducing emissions.'

At this point, I could practically hear Pauline Hanson, Malcolm Roberts and their followers screaming at their radios.

And then the interviewer asked the big, difficult question, the one I was hoping she wouldn't ask. 'Why should people care?'

It's a complicated question if you think about it. Why should we care about anything? Pure self-interest is the first, obvious, most basic answer. I care because it affects me directly. The presence or absence, the health or failure of this thing, whatever it is, will make me materially richer/poorer, stronger/weaker, healthier/sicker, happier/unhappier. At one extreme, my very survival is directly linked to its existence; at the other it could disappear tomorrow, and it would make no difference whatsoever to my life.

The trick here, of course, is having access to information I can use to make this determination for myself. What I need are clear facts, and an understanding of the causal links between the thing in question and me. If the link is clear, and the effect considerable, I'm probably going to care. If there is no clear link, or if the information I am getting suggests that the connection and the consequences are unclear, I am probably going to hedge my bets, maybe wait to see what happens before I make a decision, and certainly before I decide to change anything.

This is where science can help. Smoking causes cancer. It took a while for the science to break through and be widely adopted in public policy, but now I know. If I smoke, I have a much greater chance of dying early.

Simple. My grandfather and my mother-in-law were both big smokers; both died early from cancer (at forty-three and fifty respectively). I don't think I'm going to smoke.

Of course, if there are people benefiting from you smoking (tobacco companies), and they don't want you to stop, they can try to obscure the facts, or better yet provide you with a different set of facts, so you can tell yourself that it's okay. To keep you smoking, they can try to convince you that the science is untrustworthy. That word again. They can call the work of the world's most prestigious medical institutions and researchers 'junk science', or another favourite, 'pseudoscience'. In its place they will offer you not their own carefully researched, data-driven analysis, but a mash-up of oblique and disingenuous criticism, opinion, and ideology masquerading as science. And they do this because it works. If enough people buy into their arguments wholesale or are simply made to believe that the science is unclear, change becomes politically difficult.

So why should we care about coral reefs? Many of us will never see one (other than perhaps on TV). Coral reefs cover only about 260,000 square kilometres of the earth's surface, about 0.2 per cent of the sea floor, scattered like gems from an infinitely wealthy but careless hand across the planet's equatorial oceans. With a total area not quite the size of New Zealand and a little larger than the US state of Oregon, the world's reefs provide coastal protection, food and livelihoods for hundreds of millions of people.[2] Over 275 million people in seventy-nine countries depend on reef-associated fisheries as their major source of animal protein.[3] More than 39,000 Australians were directly employed in reef-related occupations in 2015–16, the vast majority in tourism, mostly on the GBR.[4] In comparison, Australia's oil and gas sector employed just 19,000 people in full-time, albeit highly paid, jobs that year.[5]

Globally, reefs provide direct goods and services to people valued at

US$2.7 trillion a year[6] – more than the GDP of the United Kingdom in 2020. The Great Barrier Reef alone contributes about A$6.4 billion a year in direct value to the Australian economy.[7] And all this value depends on healthy reefs. Dead, dying or degraded reefs would provide a mere fraction of the total economic benefits we now enjoy.

Another reason to care about reefs is their huge value as treasure chests of planetary biodiversity. Globally, despite their vanishingly small area, coral reefs support at least a quarter of our planet's marine species.[8] Coral reefs are the biological engines of our tropical oceans, providing food, shelter, nesting grounds and nurseries for thousands of marine species. The Great Barrier Reef is especially diverse. Snorkel or free dive off Lady Elliot Island in the southern GBR today and you will almost certainly see huge manta rays; pods of playful dolphins; exquisite lionfish; white- and black-tipped reef sharks; big spotted leopard sharks lurking in coral canyons; and inquisitive green and hawksbill turtles, ancient and unchanged across millennia. In a few minutes, you will see dozens of species of corals. These places are literally the fabric of life, an interconnected web of species that together make our planet what it is, the only living place in our universe, and certainly the only one we will ever inhabit.

But I didn't say any of that. Because self-interest may be the most obvious reason why we should care, but it is not the most important. Ask anyone who has experienced the wonder of a healthy reef first-hand, or the majesty of an untouched old-growth forest, or the pristine, sparkling beauty of an alpine glacier, about why they care, and chances are their answer will have nothing whatsoever to do with self-interest. In fact, it will be quite the opposite. We care about these places because they are beyond us, separate, timeless. They bound and transcend our short lives, and so teach us the virtue of humility. We care about them precisely because they will still be there, and still be beautiful, long after we are gone. As custodians of the planet for all of the future, we care because we must. There is no alternative.

Experiencing a thriving, healthy tropical reef first-hand is an unforgettable experience. You can still enter a secret world of hidden beauty like nothing you have ever seen anywhere else. I grew up on the west coast of Canada, where the water is cold all year round, fed by coastal glaciers, heavy with plankton and thatched with kelp, green and dark. Not like here. Here the water is warm and clear and blue.

Sinking below the waves, the first thing you notice is the detachment. You are an alien here. Your pulse spikes, your breathing accelerates. You realise that your lungs strip oxygen from air – not, like most of the life you can now see, from water. It is something you have known all your life, and yet here, it seems otherworldly. Sunlight strobes through the water in thick cloud-cut beams. The scales of fish rip colour from the sun's spectrum – every colour you have ever imagined, and some you have not. As you move, you realise that you are swimming, moving like a sea creature through this three-dimensional world. No longer confined by gravity to trudge the surface, you float over, through and across this strange topography, the living architecture of the most diverse coral reef on the planet. It is as close to flying as you can get without wings.

You take a deep breath, kick hard, dive down, equalising as you go. Here is a giant *Porites*, hundreds of years old, super tough, a survivor, its surface exquisitely patterned with tiny green polyps. Spread around it, exquisite *Acropora*, the branching corals. At a glance you can make out dozens of different forms, shapes, colours: an underwater city. Fish of every size and shape dart in and out of hidden streets and alleys, peer from the windows of protective hideouts. Some you recognise from books or programs you have seen on the TV. Everywhere you look is life.

It is destabilising. For a moment, you feel dizzy, fight the urge to get back to the surface, pull up your mask, breathe. But then, after a time, you realise that the sense of detachment you felt initially has gone. Suddenly, you are *part of it*. Some part of you, conscious or not, has made the connection. This is where we come from. So long ago that all we can feel is a Holocene echo, some sense locked deep in our hypothalamus, that we and nature are not, as we have been taught to believe, separate. We are not enemies. We only act like it.

Swim another reef, nowadays likely one further north in the GBR, or in the Caribbean or South-East Asia, and you will experience something altogether different. For the other extreme of reef health is as gut-wrenching and confronting as its opposite is exhilarating. The first thing you notice is the *silence*. A living reef is alive with sound, the trills and comic honks of fish, the crackle and snap of coral grazers, the chatter and song of dolphins. The reef platform – the shoal of carbonate rock that the coral colonies have built up over thousands of years, each generation building upon the skeletons of its predecessors – hulks before you in the sunlit water, quiet and still.

As you get closer, the damage becomes clearer. Somehow you feel like an intruder rather than an observer, a witness to unspeakable grief. It is as if a great weight has been applied from above, crushing the delicate architecture of the reef into rubble. As far as you can see, the grey bones of staghorn lie heaped and winnowed by the waves. Turf algae has started to cover over the piled skeletons of *Acropora* in a dark, slimy film. There is no colour – it is as if you are seeing everything through an old black-and-white TV set. The few fish you can see seem to be searching the wreckage for something. Food? Their favourite hiding place? They too seem to have lost their colour and brilliance, and now mourn in dull pewter, black and iron. Here and there, if you are lucky, a colony of remnant *Porites* or one of the other tough, massive coral species survives, circled forlornly by a few blue damsels, their iridescent cobalt scales a reminder of all that has been lost. You wonder where all of the other fish have gone. You emerge from the water feeling strangely lessened, as if something essential has been taken from you. And you feel something else, too, something deep in your gut, visceral. Or at least I do: I feel ashamed.

And it is this that makes the current wars so difficult to understand. There are so many reasons to care about our reefs and oceans and the natural world that we exist within, it seems impossible that any sane person could knowingly advocate for their destruction. If you care about Australia's economy, about jobs, then you probably care about the GBR and the country's other coral reefs. Senators Roberts and Hanson care about jobs on the reef. Their 2016 trip to The Keppels was in support of reef tourism operators. Members of the Great Barrier Reef Marine Park, and the scientists at the Centre of Excellence in Coral Reef Studies at JCU care about the reef, as do the people at AIMS and dozens of other scientific institutions around the world. Most Australian citizens care about the reef, and many believe it is a critical part of our national identity.[9] Millions of people around the world care enough about the reef to support its protection, and to come and visit it. Many more care because in their hearts and minds, the reef is simply too beautiful, too wondrous *not* to care. It was this that I told the interviewer.

So, if everyone cares, why are we fighting?

~

A week after the interview, I was still pondering that question. As the concern over the impacts of the 2016 and 2017 bleaching events reached further into Australian society and across the world, individuals and groups politically left, right and centre were lining up to express how much they cared about the reef. Conservation organisations raised the alarm, tying the bleaching to climate change, stressing the urgent need to reduce emissions. Greens and Labor politicians used the plight of the GBR to hammer the conservative Liberal–National coalition government on their largely ineffectual and half-hearted climate policies. The government, internally divided between right-wing pro-fossil-fuel climate-change deniers from rural seats and progressive moderates representing largely more urban constituencies, battled to find a line that would keep the coalition from rupturing. Organisations like the Institute of Public Affairs, supported by the right-leaning Murdoch press, showed their care for the reef by claiming that the GBR was just fine, thank you very much. Their logic was inescapable: if climate change was nothing more than a hoax cooked up by the enemies of progress, then it could not possibly be affecting the reef. The damage claimed by scientists was overblown, and if anything it was green hysteria that was hurting the reef by discouraging tourists from visiting.

As I walked down the hill to my office that Monday morning, with the Pacific stretched out blue and calm before me, the rocky point where shipwrecked traveller James Morrill washed up in 1846 just visible above the trees, I was struck by the viciousness of the current rhetoric, and how quickly it had escalated.

That afternoon, I was reminded by my assistant of a commitment made by my predecessor to speak at an upcoming ceremony at James Cook University, announcing the donation of a collection of rare books and papers to the university's Eddie Mabo Library. The placeholder for the event had been in my calendar since I'd started, and I had assumed that it was just another event like so many others I was expected to attend – say a few words, shake hands, represent the organisation. Compared to the flurry of press interviews and government briefings that had been

31

dominating my days so far, it seemed a lower priority issue, and I hadn't paid it much attention.

At this point, I was barely keeping up with the volume of information that was being thrown at me. My days were filled with briefings on every aspect of the institute's operations, from the status of our two large ocean-going research vessels, to the capabilities of our laboratories across the country, to the details of our science programs, finances, personnel and partner relationships. And the furore over the loss of so much of the Great Barrier Reef, so quickly, had been intensifying. The questions were coming thick and fast from every direction: How did this happen? What was causing it? How bad could it get? Was there anything we could do about it? And, inevitably, who was to blame? As the holders of the longest and most comprehensive records on the health of the reef, everyone was coming to us for answers.

On some days, I doubted that I was going to be able to cope. Much closer to the end of my career than the beginning, I was starting to understand what imposter syndrome felt like. Increasingly, this upcoming speech seemed like an unnecessary distraction. I couldn't decline, but I could make it as brief as possible, stick to the basics, and keep to the briefing notes my colleagues were preparing for me.

And then, a few days before the ceremony, the institute's librarian put a copy of a letter on my desk. It was dated 1982, addressed to Dr John Bunt, the then director of AIMS. Typed on an old manual typewriter and corrected in an unsteady hand in faded blue ink, Sir Charles Maurice Yonge, then in his eighties, described his private library, consisting of thousands of papers, letters and books collected over a lifetime of travel, exploration and scholarship. He was old, in ill health, and mourned no longer being able to do the things he once could. His body was failing, and he knew he was near the end. Looking back, he described the year he spent on the Great Barrier Reef in the 1920s, leading the first ever scientific expedition to the reef, as the best of his life. He wondered whether AIMS, where he had spent six weeks in 1978 and formed lasting friendships, might be interested in buying his personal library.

Standing at my desk at the AIMS facility in Townsville, I re-read the letter several times. I had never heard of Sir Maurice Yonge, but it wasn't often that you got to read someone else's unedited personal correspondence, with such powerful emotions put so plainly. My own father, who would die in 2020, was about the same age when I read the letter as Sir Maurice was when he'd written it. And, like my dad, Maurice had been in ill health. Here was an old man expressing the joys and sorrows of age in a way my father never would or could. Suddenly, this collection took on a new importance that surprised me.

Over the next few days, I started to dig into the records. It turned out that Sir Maurice did end up making the deal with AIMS. The agreed price of £21,000 was paid, and the library was shipped to Australia in 1983, more than a hundred boxes of it. The collection was duly catalogued by AIMS staff and, because of a lack of space in the AIMS library, stored in the windowless concrete-floor basement of the main building at AIMS's Cape Cleveland headquarters. And there it had remained, collecting mould and dust for almost forty years, a hidden treasure. Other than the occasional glance from interested staff at the institute, the library was ignored, and as the years passed, largely forgotten. And gradually, the humid, tropical conditions of northern Queensland took their toll.

And then, in 2011, AIMS began a major refit of the headquarters building. The original AIMS library was moved and downsized. Many of the books were earmarked for storage or replaced with digital archives. A new central hub was installed, and in the search for additional storage space, Yonge's old library was rediscovered, wasting away deep in the basement. Thankfully, rather than consigning it to landfill, my predecessor, Dr John Gunn, had the collection independently assessed and valued. He was shocked when the report came back a few months later. The valuers conservatively estimated that Sir Maurice's library, which included thousands of rare books, some dating back to the early 18th century, was worth over half a million dollars.

Unfortunately, however, some of the documents were already showing signs of serious deterioration, and the whole collection was at risk. Something needed to be done, and quickly. He reached out to colleagues at James Cook University. JCU had the proper facilities to house and preserve antique books at its state-of-the-art Eddie Mabo Library. Talks were opened, and a deal was struck not long after. AIMS would donate the collection to JCU, and JCU would house and care for it. The long job of recataloguing and transferring each item of the collection began.

But it was Sir Maurice's own story, his and his wife's, that intrigued me most. Over the next few days, I read as much about it as I could. As with most things, the basic information was on the internet: dates and names and places. I went to the JCU library, where the collection was already being examined and recatalogued. What struck me as I leafed through Sir Maurice's old letters and the carefully albumed black-and-white photographs from his 1927 expedition, was how much the world had changed since then. A vision of the coral core flashed in my head – all those years rendered in hard calcium carbonate. In less than a hundred years, we'd gone from a pristine reef of unrivalled beauty, diversity and abundance, to facing the shock realisation that the warnings of more than three decades were coming true – the reef was in big trouble.

How exactly did we get here? Perhaps the answers to our present conflict lay, in part at least, somewhere in the past.

3

Yonge and the Early Days
of Exploration

The answer to the question 'How did we get here?' starts at the dawn of the last century. In 1927, in the wake of the most destructive war ever known, and with the dark shadow of the Spanish Influenza just lifted from the world, a young University of Edinburgh zoologist left the United Kingdom with his new bride for the trip of a lifetime. Maurice and Mattie Yonge were part of the first scientific expedition to the Great Barrier Reef, the almost mythical system of coral reefs and islands strung out along the coast of Queensland in the north-east of Australia, which Captain Cook had discovered 150 years before when he ran aground on a reef just off what is now Cooktown.

Steaming from Portsmouth aboard the SS *Cumbria*, the Yonges transited the Mediterranean, passed through the Suez Canal and the Red Sea, before heading out across the seemingly endless expanse of the Indian Ocean, bound for Australia. A photograph of the party boarding the liner in England shows them excited and happy, young faces full of that suspense that comes when setting out on an amazing adventure.

Maurice was just twenty-seven at the time,* Mattie was barely twenty-three. I imagine them standing at the railing as the ship transits the Red Sea through to the Gulf of Aden, the volcanic peak of the port city that gives the gulf its name rising up on their port beam, the sea breeze in their hair, the sun rising over British Africa, the whole of their lives stretched out before them – an ocean of time.[1] Dolphins, dozens of them, ride the ship's big bow wave, dancing in and out of the clear blue water. Big shearwaters and long-range petrels arc and spin in the azure and gold sky.

The tumultuous events of the rest of the 20th century, the scientific, industrial and political developments that would shape the fate of the earth, had not yet occurred, and to most were beyond imagining. The speculation-driven stock market crash that triggered the Great Depression would come before the year was out. The Second World War was over a decade away and would claim the life of at least one member of the expedition party. The atomic bomb was still only a distant theoretical possibility. And the notion that humankind could fundamentally transform the planet's climate, and thus endanger the very wonder they were off to explore, was nothing but a theory limited to a few arcane papers published in obscure scientific journals.**

In the 1920s, the earth was a place of relatively few people (about two billion at the time; it had taken us over 125 years to double our numbers from one billion), and it was still graced by vast uncharted territories and uncatalogued ecosystems, teeming with life. Our oceans remained

---

* The same age as Charles Darwin was when he arrived at the Cocos (Keeling) Islands on the HMS *Beagle* eighty-one years before.

** In 1872, John Tyndall, the Irish physicist, published *Contributions to Molecular Physics in the Domain of Radiant Heat*, in which he described the heat-trapping properties of gases such as carbon dioxide (CO_2), water vapour (H_2O) and methane (CH_4). Svante Arrhenius, the Swedish physical chemist, published a paper in 1896 connecting CO_2 concentrations in the atmosphere to global warming.

places of mystery and danger. Thousands of ships and sailors were lost at sea every year, run aground, sunk, broken by cyclones and disappearing never to be seen again. Fishing was still a difficult occupation, powered mostly by human muscle.

The Australia that greeted the Yonges when they disembarked in Brisbane in 1928 was still young in the European sense, despite having come of age in the bloodbath of the First World War. Construction of the country's first parliament house in Canberra was completed that year. Charles Kingsford Smith and his copilot Charles Ulm amazed the country with the first ever flight around Australia, leaving Sydney and flying north via Brisbane, then striking west for Longreach, Darwin, Carnarvon and Perth, and then back via Adelaide and Melbourne, overflying much of Australia's 34,000-kilometre coastline. The country's census that year put the population at just over six million, an increase of 120,000-odd over the previous year. The census only included people of European descent. Despite being home to the world's oldest human civilisation, the country where Maurice and Mattie Yonge had just set foot did not count the First Australians as citizens.

Waves from three of the world's great oceans lap Australia's shores. The warm Pacific, home to the Great Barrier Reef, was the Yonges' 1927 destination. At the time, the GBR was home to countless millions of fish and sharks and rays; dugongs thrived in prairies of sea grass; and turtles came ashore at night to nest on pristine beaches. Further south, off the coast of New South Wales, lay the world's largest giant kelp beds. The Southern Ocean stretched uninterrupted from Cape Grim on the southernmost point of Tasmania all the way to Antarctica, cold and clear and teeming with life. Southern right whales still braved the whaling boats and congregated in their hundreds in the waters off Albany in Western Australia and in the isolated expanses of the Great Australian Bight. In the west, the Indian Ocean ebbed and flowed, languid and warm, through the embayments of the Kimberley and over the isolated coral reefs of the Rowley Shoals.

Back then, most of these places remained in a condition not vastly different than they had been through tens of thousands of years of Aboriginal stewardship, and for much longer before. Their waters were clean and unpolluted. Seasons kept to their ancient rhythms, sure and dependable. And in a last few remote Indigenous communities, life still carried on in the traditional way, in balance with the sea and its creatures.

Today the northern Queensland coast and the GBR is still a stunningly spectacular place of sublime reefs and isolated coral keys studding an aquamarine and cobalt sea. Snorkelling or diving one of the better-preserved reefs of today, such as Lady Elliot Island in the south of the GBR, or one of the isolated reefs off the coast of Western Australia, would surely come close to what it was like back then.* But so much has changed almost beyond comprehension.

The Yonges' journey from England by steamship and railway, by sailboat and car over rough dirt tracks, took over eight months. I had first arrived in Queensland from Western Australia in less than twenty-four hours, catapulted through the stratosphere, whisked in air-conditioned comfort along paved roads to the front gates of the institute. Maurice and Mattie and their team lived in a makeshift camp on an isolated island, and supplies were delivered by sailing barque every few weeks. We had a comfortable house overlooking the Pacific and could drive into town whenever we liked to browse aisles bursting with every imaginable commodity.

---

* At the time of writing, both these Australian reefs remain in excellent condition. Unfortunately, there is an ever-increasing likelihood that this may not be the case by the time you read this book.

Yonge's journals spoke of virgin landscapes, coral reefs thriving in sparkling clear waters, unharvested and unfished, teeming with thousands of reef fish, giant clams hundreds of years old, ancient tritons and so many turtles they could not be counted. The few roads along the coast were still mostly unpaved, and life in the small communities that dotted the coast was hard.

When I drove through Townsville, I could see shopping malls and car dealerships and the stick-like cranes of the port in the distance, and on the outskirts heading south, the circling birds and man-made hills of the city dump. The curve and headlands of the beach Heidi and I walked most evenings haven't changed much physically since Yonge was here, but now they are littered with plastic. We would take carry-bags each time we went to haul off as much as we could carry – plastic bottles, flip-flops, lost and tangled fishing lines and lures, wrecked plastic chairs, wrappers with Korean and Chinese labels and barcodes, toothbrushes and toys, bottle caps and baseball caps. The sheer volume of crap was shocking. The few turtles that still came up to nest on our beach were now much more likely to have their eggs plundered and eaten by feral pigs within days of laying than have them grow to hatchlings and return to the ocean.

Yonge's GBR of the 1920s, in contrast, remained largely untouched, unexplored and poorly understood. We have only a few black-and-white photographs and the detailed records of Yonge's journals, alongside those of a handful of other observers from that time, to give us a glimpse of what it must have been like. These shifting baselines – the difference between what it was and how it is now – are hard for us to notice day by day. The changes come slowly and accumulate over time. But it is only over decades and generations that the true extent of such change can be measured. Without first-hand experience, unless we see something for ourselves, we have difficulty accepting it as fact. Could it really have changed that much?

Sir Maurice made the trip from the United Kingdom in 1927 to study a natural marvel. I was here, as CEO of one of Australia's premier marine research institutions, to help save it from destruction. But to save something, you need to understand not only the threats it faces now, but how it came to be in danger. What was it really like back then, almost a century ago? Had our oceans and reefs really changed that much in just a few generations? And if so, how had it happened?

~

For Maurice and Mattie Yonge, the journey to northern Queensland actually began in England in 1922, with the establishment of the Australian Great Barrier Reef Committee. Two of its founding members, Sir Matthew Nathan and Professor Henry Richards, suggested a scientific study of the reef to investigate, among other things, the theories on coral atoll formation put forward by Charles Darwin. By 1927, a team of twelve scientists had been identified, including Maurice Yonge and Mattie, who would act as medical officer for the expedition. The core team included two other married couples: the marine naturalist Frederick Russell and his wife, Gweneth; and University of London zoologist Dr Thomas Stephenson and his wife, Anne. Also among the team was a botanist from the Natural History Museum; Dr Andrew Orr and Dr Sheina Marshall, naturalists from Millport Marine Laboratory in Scotland; James Steers, a geomorphologist and surveyor; and his assistant Michael Spender, who would go on to become a squadron leader in the RAF in the Second World War and be killed in action days before the war's end. Other than Frank Moorhouse, a marine biologist from the University of Queensland, who was to provide local zoological knowledge, and AG Nichols of Perth, who acted as Yonge's assistant, the expedition was very much a British affair.

One of the Yonge expedition's key objectives was to verify Charles Darwin's theories on coral reef formation, promulgated almost a century earlier. The fundamentals of coral reproduction, another of the questions the expedition sought to investigate, would remain a mystery for another half-century again, until the early 1980s, when a group of early-career scientists at James Cook University first described mass coral spawning on the GBR.[2] The other stated goals of the expedition were to collect hydrographic and meteorological data, observe and catalogue the coral reefs in the vicinity, and study the geology of the atoll and nearby islands. There was also an economic component to the expedition. The team was to assess the viability of trochus shell farming, and the commercial potential of fish, oyster and turtle populations in the area.

In July of 1928, after a journey of 15,000 miles, and with Yonge newly appointed as the leader of the expedition, the team arrived in northern Queensland. Yonge's *A Year on the Great Barrier Reef: The Story of Corals and of the Greatest of their Creations* – his bestselling book, published in 1930 – recounts his party's time on just one of the reef's more than 600 islands.[3]

Low Island, where his expedition made their camp, is located about twenty kilometres from the town of Port Douglas on the Far North Queensland coast. Today Port Douglas is an upmarket tourist town, full of high-end resorts and boutiques catering for the thousands of dive and snorkel tourists who flock there every year to see the reef. But back then it was an ex-mining town in severe decline, with a little-used railway station and a courthouse serving a district whose turn-of-the-century population had been estimated at about six thousand but had dwindled considerably since. Yonge's black-and-white photograph taken from the ridge above town shows a few buildings scattered along the fringes of the estuary; 'more of a memory than a reality these days', as he put it.

Conditions for the team on Low Island were difficult. As Yonge described it, 'The cay was no more than a mound of sand, only six feet above the high-water mark, oval in shape, about 185 yards long and 110 yards wide, with a total area of some three-and-a-half acres.' Their only company was the three-man crew of a tin-clad timber-framed lighthouse, built in 1878. Fresh water, while plentiful during the monsoon season (when keeping dry was the main problem), was non-existent the rest of the year. Food, water and other supplies had to come by sailing barque from the mainland. Yonge wrote of the party's delight at receiving mail and a newspaper every week or so, 'bringing the realisation that the outside world *was* still there and getting on quite all right in spite of our temporary absence'. Yonge described the monsoon season between December and April as living in a 'perpetual Turkish bath', adding that 'such conditions do not make for efficiency of mind or body'. Sandflies were a constant nuisance. Sunburn was a problem that the fair-skinned Britons had not anticipated, and which affected certain members of the party quite brutally. During the dry season (between May and November, when the trade winds blow strong and dependably on the reef for months on end) the party suffered unexpectedly from cold. Geared up for a tropical sojourn, the shorts and singlets they wore almost exclusively left them shivering and miserable.

As time went on, they adapted. Their research station took shape, with a kitchen, married and single quarters, and eventually a cosy library with a small heater. They even fashioned a primitive refrigerator using a rudimentary seawater heat exchange system, and could enjoy 'iced drinks in mid-summer on a lonely coral island sixteen degrees south of the Equator'. Scientific instruments such as tide gauges, thermometers, barometers and wind gauges, were begged and borrowed from mainland institutions, including the Commonwealth Meteorological Bureau. The whole expedition was put together on a shoestring budget, and a frayed one at that. They were understaffed for the tasks they had set themselves, and despite a steady stream of volunteers from the mainland who lent a hand, they worked long hours and well into most nights, propelled by their passion for the science, and a belief in the importance of what they were doing. 'It was not until after I left the island,' wrote Yonge, 'that I realised fully under what a strain we had been living for many months past.'

The scientific work began the day they arrived. Days were spent collecting, describing and cataloguing samples, surveying the island and its surroundings, and recording atmospheric and oceanographic data. Luck blessed them early, in the form of the *Luana*, a 39-foot ketch owned and piloted by a Mr AC Wishart of Brisbane, who, after ferrying up some volunteers, offered to stay on, providing the expedition with its very own research vessel. The *Luana* allowed the team not only to access surrounding reefs and islands, but eventually to sail as far south as the Capricorn and Bunker groups and as far north as Mer Island in the Torres Strait. Everything was new, waiting to be described and investigated. There was literally no end to what needed doing, and Yonge's book is full of references to how little time they had, and how quickly, in the end, that thirteen months passed.

~

Yonge's expedition turned out to be a great success. The newspapers reported widely on the team's progress, generating huge public interest in Australia and back in the United Kingdom. At that time, most people's understanding of coral reefs was limited to the danger they posed to navigation. Stories of shipwrecks were commonplace still and, then as now, stories of disaster involving sudden loss of life tended to dominate the headlines. The majority of Britons, and indeed most Australians, had never seen a coral reef, and those who were aware of their existence had little idea of the sheer scale, vibrancy or eye-popping splendour of the GBR.

Photographs and descriptions of the reef produced by the expedition and various journalists who joined the team for various periods kindled the public's imagination. The team published seven full volumes of scientific material and authored several academic journal papers based on their work at Low Island. In 1930, Yonge published his memoir of that time. Such was the success of his book and the widespread news coverage that tourists began flocking to the Low Isles to collect coral specimens and the beautiful shells described by Yonge and his colleagues. Within a few years, the reef flats surrounding Low Island were swept clean. The reefs of the GBR would have to wait another forty years to become protected.

The wonder and excitement of that time comes through in every page of Yonge's memoir. The sense of time's inexorable flow is palpable. The book I read was a ninety-year-old first edition from the JCU Library collection. As I read, it weighed heavy in my hands like an archaeological relic. The frayed edges of hand-drawn maps protruded from the fore-edge of the text block like afterthoughts. Members of the expedition, long dead, stared out of the pages from faded black-and-white photographs into an unknown future. When Yonge wrote it, my father had yet to be born. Now they are both dead.

There was romance there too: the romance of another time, when the world was still pure and there to be discovered by those lucky enough to be able to travel, and intrepid enough to suffer the privations and risks such adventure entailed. I read the book in a few days, sitting on our veranda overlooking the ever-changing waters of the reef, reflecting on the gulf that separated me from Yonge, then from now. A distance not only of time, but of technology, of knowledge and history, and of something else that I couldn't quite put my finger on. Was it awareness, perhaps?

Are we more aware now of the beauty of the world, its fragility, and our responsibility for its protection? Surely this must be the case. Our knowledge of the natural world is so much greater today than it was a century ago. We collect as much data on our oceans in a single day now as we did in a decade back then. The imagery we produce now is of incomparably higher quality and quantity than the few black-and-white photographs captured in Yonge's reports. High-quality documentaries such as the BBC's *Blue Planet* stream through the internet to the world, attracting millions of viewers. But the more I thought about it, the more I realised that my frame of reference was incomplete, hopelessly skewed, and that in fact I was asking the wrong question, of the wrong people.

For there was once a group of people who were intimately aware of the beauty of the world, whose way of life enshrined a deep respect for nature and country, and whose very existence was based on taking responsibility for the future. These were, of course, the First Nations, the Aboriginal and Torres Strait Islander peoples. But by the time the Yonge expedition arrived in northern Queensland, Aboriginal communities were already in the process of being fundamentally transformed. Many had disappeared altogether.

In his 1930 memoir, Maurice Yonge describes a visit to Yarrabah Mission on Cape York. He wrote:

> In their natural state the Aboriginals who were a purely hunting people
> in a land subject to great droughts and never well stocked with game,
> restricted the natural increase of the population by methods known
> only to themselves. This was the wise measure of a natural people, for

the means of sustenance were too slight to support a large population. Now the missionaries are gradually persuading the natives to abandon these practices, with the result that the population is increasing ... by leaps and bounds.[4]

'Gradual persuasion' was not always the favoured means of getting the natives to do what Europeans wanted. The last Woppaburra people, for example – eighteen men, women and children – were forcibly removed from their ancestral homes on The Keppels – those same islands visited by the Senators in 2017 – in 1911, just a few years before Yonge arrived.

As with so much First Nations history, many of the facts about The Keppels and the fate of the Woppaburra are disputed. Needless to say, most of the early accounts were written by the colonisers. Archaeological evidence dates Indigenous occupation of the islands back at least 5000 years, and points to an enduring connection with the sea. Evidence suggests a small population of perhaps two hundred individuals quite distinct from mainland Aboriginal tribes.[5] They had their own language and customs, a diet dominated by fish, and considered themselves very much a saltwater people.

The earliest written records suggest that the first visit by whites to the islands in 1865 resulted in calamity. At least seven members of the Woppaburra were massacred. Soon after, pastoral leases were granted on the islands and thousands of sheep were introduced. This led to several decades of conflict and the deportation of increasing numbers of islanders to the mainland, where, predictably, they suffered greatly.[6] Shunned by other Aboriginal tribes, abused and exploited by white farmers, some tried to swim back to their homeland. Several were killed in the attempt, drowned or eaten by sharks. Descendants speak of the fate of those who remained on the islands, treated as slaves, chained to ploughs, shot, and the women carried off never to be seen again.[7]

In 1911, when the final eviction occurred, the Southern Protector of Queensland's Aboriginal Peoples, Archibald Meston, argued that it was for the protection of the remaining Woppaburra people. Within a year, all but eight of the surviving Woppaburra were dead. The survivors dispersed along the coast. Today, some three hundred Australians trace their lineage to the Woppaburra people of the Keppel Islands.

~

Yonge's expedition, and the worldwide attention it brought, sparked a desire to know more about coral reefs, and about the oceans in general. But it wasn't until 1967 that the next major scientific expedition to the GBR took place, this time sponsored by the University of Liège in Belgium, with support from the Belgian Government. The team, led by Professor Albert Distèche, spent seven months in the vicinity of Lady Musgrave Island and Lizard Island, based aboard an ex-Royal Navy warship converted for scientific duty. The expedition included a number of important guests, including the famous Australian underwater cinematographer Ron Taylor and, in direct tribute to the 1928 expedition, Sir Maurice Yonge himself, who by this time had been knighted by Queen Elizabeth II.

The early years of underwater exploration were a time of fascinating individual characters, and Ron Taylor was no exception. By the time he was invited to join the Belgian expedition to the GBR, he had already developed a reputation as an underwater cinematographer par excellence. In the 1970s Taylor filmed the live underwater shark sequences for the hit Hollywood movies *Jaws* and *Jaws 2*. In 1979, he filmed the underwater scenes for the big-budget movie *The Blue Lagoon*.

In 1968, the Belgian expedition had one specific advantage of which Yonge and his colleagues in the 1920s could only have dreamed. Invented by the French oceanographer and explorer Jacques-Yves Cousteau in 1943, the aqualung, the predecessor of modern scuba systems, provided for the first time the ability to navigate below the surface untethered and unsupported for hours at a time. Finally, people could experience and observe the undersea world without relying on surface apparatus. It unlocked a golden age of marine exploration and study. In 1953, Cousteau published his first major book on ocean exploration. *The Silent World: A Story of Undersea Discovery and Adventure* was a huge international hit.

About the same time that I was discovering my first coral reef as a young boy in the Caribbean, a man now regarded by many as the father of modern coral taxonomy – John Edward Norwood Veron – was beginning his post-doctoral research at James Cook University in Queensland, focusing on corals. Charlie, as everyone calls him, was typical of the next generation of marine explorers in Australia born shortly after the Second World War. Fiercely independent, a self-described maverick, he took to the new sport of scuba diving while still at university. Within a couple of years, he became the first scientist hired by the newly established Australian Institute of Marine Science, eventually going on to serve as its chief scientist. Over the next two decades he travelled the world describing and cataloguing the immense undiscovered biodiversity of the planet's coral reefs. He went on to identify and name over a fifth of all known coral species on the planet.

It turns out that it was Charlie, while working at AIMS, who struck up a long-term friendship with Maurice Yonge, and who, behind the scenes, brokered the deal to bring Yonge's collection to Australia. When I first met Charlie, not long after I started at AIMS, he described with unabashed glee those early years of exploration. Well into his seventies, he still looked like the prototypical marine explorer of that era. Piercing blue eyes gazed out at you with a slightly boyish air. His face was weathered and craggy from a lifetime of sun and sea, and every once in a while he would flash a slightly sardonic smile that suggested not only a mischievous sense of humour, but a stubbornness shared by many single-minded pioneers. Among the first generation of scuba divers, they were the first scientists to be able roam freely beneath the waves.

'Everything on the GBR was new,' he told me. 'A whole undiscovered world. We were the first to identify the Queensland trough – the western wall, the place where the reef ends. We were diving the reef face, and it just went down and down to the abyss. No one had ever seen that before.'

Until then, the discipline of coral taxonomy had been dominated by the museums, using coral specimens collected in shallow water. Charlie was the first person to realise that individual species of coral radically changed form with depth. This now-universally accepted conclusion was met with hostile opposition at the time.

'In the seventies, very few marine scientists were divers,' he recalled. 'It's completely different now of course. Only wild young men were divers back then.' His tone took a turn to the wistful. 'It was an amazing time, those eight years or so. I dived every part of the GBR, north to south. I was in the right place at the right time. I've been incredibly lucky in my career.' He paused a moment, and I imagined he was reliving some of those moments. 'But you know,' he said, his voice cracking, 'there is so much I regret. I regret not recording the number of sharks I saw on my dives, building a record of it. When I dive now, it's completely different. There were a hundred times more sharks back then.'

It is a theme he came back to repeatedly as we spoke. Like Yonge decades before, he reflected on the palpable sense of time passing, manifested not only in his own changing attitudes, but in the physical condition of the reef itself.

4

Truth and Denial

In the end, my speech about the Yonge Collection didn't go quite as planned. I started on script with Yonge's letter to my predecessor, seeking a permanent home for his collection, and its subsequent purchase and burial in the basement at AIMS. But then I started to stray. I spoke of Yonge's wistful reflections on that best year of his life, and how much his letter had affected me personally. I channelled Charlie Veron's emotion-charged words and asked the audience to consider how much things had changed over their own lifetimes. Clearly, given what was happening on the reef, we had to change course. We needed to urgently address the challenges facing our oceans, threats which had largely arisen since Yonge's expedition: pervasive pollution, dangerous over-harvesting, and now irreversible climate change. As I spoke, I could see the expressions on the faces of the audience change. A few people were nodding. Others narrowed their eyes or stared into their phones. By the time I started talking about healing our relationship with the people who had lived in balance with the oceans for millennia, the audience, most of whom had come expecting to look at old books, was fidgeting, restless.

By the time I saw my host signalling me – slicing the blade of her hand back and forth across her neck – I realised I had gone over time, something I rarely do. I wrapped up with the obligatory thankyous and sat back down, a little embarrassed, to some scattered applause. Even with what was going on out on the reef, this was still uncomfortable territory for so many here in Queensland, where coal is big business, and the rate of land clearing for farming and mining matches that seen in the Amazon. It is the essential dilemma at the centre of this war.

The tourism industry in Queensland thrives because people from all over the world, millions of them, want to come and see the most spectacular reef system on the planet. Those visitors support not just the boat tour operators, but the hoteliers and restaurateurs and laundries and car rental companies and food wholesalers and gift shops who collectively employ over 65,000 people full-time. Before the Covid pandemic, tourism was the single largest contributor to Australia's blue economy – bigger even than offshore oil and gas – with the GBR the single largest destination by value.

Ostensibly, this was part of the reason why the Queensland senators were so averse to any bad news about the current and future health of the reef – if the reef was dead or dying, then people wouldn't spend money to come and see it. It made sense. Except for the fact that another big contributor to Queensland's economy is the coal sector. The Queensland Government expects to net over $20 billion in coal royalties in 2025–26. Over 26,000 people in that state worked in the coal sector in 2021. And if the biggest threat to the future of the reef, and therefore the tourism industry, is climate change, and coal is the biggest single contributor to climate change, then we clearly have a problem.

Sandwiched in the middle, often unfairly, are the state's farmers and graziers. In 2022–23, primary industries contributed over $23 billion in value to Queensland's economy, employing over 70,000 people,[1] with agriculture the biggest component by value. But farming and grazing

52

have also contributed to the deteriorating water quality that the reef has experienced since European settlement.[2] Clearing of native forests for pasture has resulted in erosion of land, carrying sediment into rivers and eventually out to the reef. Sediment can block the light that corals need to survive, and smother them on the sea floor. Application of fertilisers and pesticides for crops such as sugar cane, widely cultivated along the North Queensland coast, can also harm corals.

As the notion that the reef is in trouble gains traction with the public, the media, and with politicians, each of these sectors, and the people whose current and future livelihoods are directly tied to their ongoing success, must vie to get their message out. Each makes the case that it is a valuable contributor to the state's prosperity. Each tries to gild its environmental credentials and, in its own way, to downplay the threat to the reef and its contribution to that threat. The political high-wire act between these opposing realities has consumed Australian and Queensland state politics for over two decades.

Do we want a healthy reef, or jobs and economic growth? To some, these two goals are compatible, or can at least be claimed to be compatible. To those who sit on the extremes of the fight, they are fundamentally opposed – we can have one or the other, but not both, and we are all going to have to choose which side we are on. Many others, confused by the increasingly strident shouts from the extremists, just walk away from the whole thing. It's simply too hard and stressful – and anyway, if it really is a problem, if the reef really is at risk, then surely the government will look after it. And what of the aspirations of the traditional owners and custodians of these places? In the audience at an event like the one I was at, you could be sure that there were people from all sides of the argument.

When the other speeches were done, I circulated with the guests awhile, posed for a few photos and did a quick TV news interview.

Then I slipped away and headed back to work.

In the car on the way to Cape Cleveland, I reflected on what I had said, and the emotions that had pushed me off course. I determined that in future I would try to keep myself in better check. Throughout university and my early apprenticeship, I had always been taught that science and emotion didn't mix. Our role as scientists was to remain dispassionate and report on the facts. It was up to others, decision-makers and those who made the laws, to turn the science we provided into policy. The standard line was that once scientists became advocates, they ceased to be objective. And yet, I was passionate about the work I did, and I knew that scientists like Charlie Veron and Ove Hoegh-Guldberg and Britta Schaffelke, who had spent their lives studying the reef, were powerfully invested in what they did. It was why they and so many others worked long hours and spent months every year at sea away from their families. For many of them, that passion was why they chose marine science as a career in the first place – they loved and cared about the oceans. It was a constant balancing act: remaining objective, keeping the passion in check, while being true to yourself and what motivated you.

But it was the things I didn't say that day that I regretted most. I was a newcomer to the Great Barrier Reef, a century after Yonge. But already I could see just how far we had strayed from the balanced relationship with the sea that had allowed Aboriginal people to live here for thousands of years and still bequeath to people in Maurice Yonge's time a reef as healthy and vibrant as it had been when Europeans were erecting Stonehenge.

How did it happen, this lessening? Was it always just an inevitable consequence of human progress, or were there points along the way where we could have made different decisions? And more importantly, could we learn from the consequences of those decisions, now that the situation was so much more urgent? I was determined to find out.

~

A few days after the speech I arrived at work to find my email inbox sprinkled with messages from addresses I did not recognise. As the head of a public research agency, I made it a point to answer any queries from the general public directly, personally and quickly. I opened the first email. It contained a single sentence: *You talk shit about the reef.*

I deleted it and move on.

The next email contained a long, detailed and heartfelt argument against the science of climate change. Written by a retired engineer, the missive covered most of the now-familiar climate denialist tropes readily available on the internet: the Little Ice Age; the (incorrect) assertion that increasing temperatures drive rising atmospheric CO_2 concentrations (rather than the other way around); a stab at the inaccuracy of climate models; a reference to the manufactured climate-gate controversy in 2009, involving emails between researchers at the University of East Anglia; pointing out the susceptibility of temperature records to error; and railing against the failings of renewable energy. The tone was polite and respectful, but the content was the usual mash-up of unsubstantiated assertions, conspiracy theories, accurate fact used to support irrelevant points, and long-debunked clichés sourced from the dank corners of the web.

I thanked the gentleman for reaching out, provided links to a range of peer-reviewed journal papers from top scientists that set out the data clearly linking emissions of greenhouse gases to rising global temperatures. I also attached a few papers describing how the radiative forcing caused by these gases was driving long-term changes in rainfall patterns, raising sea levels and increasing the frequency and severity of extreme weather events, and damaging the health of ecosystems. I also pointed out that while I agreed that certain types of renewable energy did have limitations, and could appear expensive, these things had absolutely nothing to do with whether human-driven climate change was real or not.

I hit send, felt good about it. We all had to keep talking about this, trying to find common ground, trying to learn. I liked the way this man had reached out and prompted a civilised discussion. Hopefully he would feel the same way about my response.

But the next email took my breath away. It felt as if I had been kicked in the chest, and I took an involuntary step back from the screen. The language was foul – the worst – and the physical threats to me and my family were graphic. I took a deep breath and hit the delete key, then tried to get back to work. But the effect resonated, and suddenly found myself thinking of another quite different but equally threatening email I had received a decade earlier while still working in the private sector.

As director of sustainability for what was then called WorleyParsons, the big ASX-listed engineering company whose client list is dominated by large energy and mining companies, I had written to a number of colleagues suggesting that the upcoming Australian tour by leading climate denier Christopher Monckton, the bug-eyed 3rd Viscount Monckton of Brenchley, hereditary peer, self-styled member of the United Kingdom's House of Lords,* and Conservative Party political adviser, was something to be avoided.

Within days, my email had been copied and forwarded across the country, coming to the attention of several Australian climate denialist internet warriors, including the outspoken blogger Jo Nova. The response was immediate and vicious. Over the next few days, I received literally

---

* Monckton has referred to himself as a member of the House of Lords, the upper legislative body of the United Kingdom. The House of Lords authorities have asserted that Monckton is not and never has been a member of the House of Lords.

hundreds of emails, most of them containing nothing more than the best and most-loved Anglo-Saxon swear words.

What came back to me now, as I stood and stared at the screen, was the email I received more than a decade ago from a senior trader in one of the big financial houses in Sydney. It was addressed to the WorleyParsons CEO, the affable, intelligent and reserved John Grill, and cc'd to countless other investment houses around the world. The email pointed out that if a senior director of the company actively supported climate action, thus biting the hand that fed it, then investors worldwide should immediately put out sell orders on WorleyParsons stock. Within minutes my boss was on the phone expressing his displeasure and counselling discretion, whatever my personal views about the issue. My offer to resign was rejected, but I was left chastened and bewildered by the savagery of the attacks.

Home that afternoon, back in Queensland, I found my personal email and Facebook account strewn with similar attacks. Many, perhaps not surprisingly, appeared to be from the United States. I deleted, reported and blocked. Not much point in doing anything else. Seemed not much had changed in the last ten years.

That evening, Heidi and I climbed up into the granite hills behind our house. After a while we emerged at the top of a massive rounded outcrop with views in every direction, the rock ancient and weathered. Below us, Bowling Green Bay was in flood, the tidal flats sheened in silt-turbid water under shoals of grey cloud. We watched the waves reach the edge of the mangroves, then slowly infiltrate the forest. A flock of sulphur-crested cockatoos took wing and moved noisily inland towards their evening roosting site. The white birds glowed tangerine in the setting sun. The AIMS site was quiet below us, dark in the shadow of the hills, just visible through the trees, the staff gone home and the parking lots empty. Our conversation went something like this:

Me: It's beautiful here.

Heidi: Yes.

Me: I got another death threat today.

Heidi: [silence, and then] What this time?

Me: Don't worry. Just the same old shit. People who do this kind of thing, meet them in a dark alley, they'd run so fast back to the shadows you'd never even get the chance to square up. They never even use their real names. It's just bullshit.

Heidi: It makes me sad. We need people to be brave.

Me: It's as if we've learned nothing in the last ten years.

Heidi: More like twenty or thirty. We're still clearing native old growth forests, even here in Australia, right here in Queensland.

Me: They've figured out that if climate change doesn't exist, and we're saying that climate change is what's causing this bleaching, killing all this coral, then they have a problem. They've attacked every other part of the climate science for decades and it's pretty much worked. So now it's our turn.

Heidi: [taking my hand] Are you okay?

Me: Yeah, I'm okay. It's not about the reef. This is just a proxy war to these people. They are using the reef to attack climate change. And they're using the Karl Rove playbook: intimidate people who know what's going on, and a lot of them will just keep quiet. Sow doubt in the minds of the public, and most won't know what to think.

Heidi: And then nobody will do anything.

Me: Exactly. That's all they need. Inaction. Silence.

Heidi: And in the end, everyone loses. All of us.

A few days later I received a call from our communications manager. I was in Perth, working for a week with our team there. An op-ed had appeared overnight in *The Australian* newspaper, authored by a physicist from James Cook University named Peter Ridd. In it, he questioned whether the reef had in fact suffered any unusual damage at all.

I had first heard about Peter Ridd in 2016. During the severe bleaching events of that year, Sky News had run stories playing down, or even questioning the veracity of the survey data and of reef science altogether, featuring opinion from Ridd.[*][3] For more than a decade Ridd had been claiming that climate change was a hoax, that the GBR was just fine, and that the public was being deceived by the green side of politics and grant-seeking 'intellectually bankrupt'[4] scientists into believing that it was under threat.

A long-time researcher at JCU, Ridd had done some of the earliest work on measuring sediment flows to the reef. He became an epigone of another James Cook University professor, Bob Carter, who by this time had become one of the rockstars of the international climate denial movement and a regular fixture at climate denial conferences worldwide. Over time, Ridd had increasingly expressed the view publicly that other reef researchers were producing poor quality work.[5] In 2005, he joined the deceptively named Australian Environment Foundation, which was formed by right-wing think tank the Institute of Public Affairs for the express purpose of countering what they called environmental fundamentalism,[6] particularly around climate change.

The IPA itself was funded by some of Australia's richest people, including mining heiress Gina Rinehart and News Corp owner Rupert Murdoch, and had close ties to Australia's conservative Liberal Party.[7] Its website hailed it as 'the voice of freedom', espousing a list of policies to

---

[*] In this interview with Sky News on 16 April 2023, Sky's commentators stated that 'no one knows more about the reef than this guy'. After quizzing him on crown-of-thorns starfish, they spent the last third of the interview asking him about nuclear energy.

'fix Australia' that included withdrawing from the Paris climate accords, privatising the ABC and the SBS, reducing corporate taxes, repealing the *Fair Work Act* and establishing a royal commission into the Bureau of Meteorology's 'tampering with temperature data'. Their stance was consistently pro-fossil fuels, anti-public broadcasting, and dismissive of the science of climate change.[8]

Back in 2007, in a piece called 'The Great Great Barrier Reef Swindle', originally posted on the website of fellow climate change denier Jennifer Marohasy, Ridd put his cards on the table:

> So why have we been swindled into believing this almost pristine system is just about to roll over and die when it shows so few signs of stress. There are many reasons and processes that have caused this and some of them are the same as why we should all be more than a little sceptical about the hypothesis that CO_2 is causing global warming.
>
> The first reason is that there is some very bad science around ... we have many organisations and scientists that rely for funding on there being a problem with the GBR. Most grant applications on the GBR will mention at some stage that a motivation for the work is the threat to which it is exposed. I confess that I do this in all my applications – it's the way the game works.[9]

In an opinion piece published in 2008, Ridd made his political views clear, describing the modern green movement as 'hypocritical, illogical and treacherous ... hijacked by quasi-religious ideology'.[10]

As the 2016 bleaching event was unfolding on the reef, Ridd publicly accused his JCU colleague Terry Hughes and the GBR Marine Park Authority (GBRMPA) of misrepresenting photographs of the reefs around Stone Island, near Bowen, to mislead the public about the state of the reef and the effects of climate change. The response from Hughes and the university was swift. Ridd was accused of 'failing to act in a collegial way and in the academic spirit of the institution', so breaching JCU's code of conduct. Ridd countered that he was doing nothing more than pointing out flaws in the way Hughes and the Parks Authority were presenting information.[11]

Ridd ended up being formally censured by the university, and soon after went public with details of his censure. In an interview on Sky News in August 2016, he bared all, accusing the university of violating the principles of academic freedom and freedom of speech. He again attacked other reef scientists, claiming that their work was untrustworthy. These were the opening salvos in a dispute between Ridd and JCU that would unfold over the next five years, and which would eventually see him dismissed from JCU.[12]

On 28 November 2016, Senator Malcolm Roberts, fresh from his trip to the reef, brought a motion in the Senate for a debate on a matter of public importance. The topic: 'The policy position of the Australian Government towards the continuing robust health of the Great Barrier Reef and the threat of environmental alarmism'.

The debate received enough votes to proceed, and Senator Roberts rose from his seat to launch proceedings. 'The onus is on honourable senators gathered here to expose the truth about our reef and repudiate the lies being told by extreme left-wingers hell-bent on control.' He went on to describe his recent visit to the reef, and how good it had looked. Then he got down to business. 'It is an affront to Queensland that some Australians want to destroy the reputation of our most prominent icon ... through fake science and lies,' he said. 'The destruction of the reef's reputation is being done to buttress the Greens' agenda of falsely claiming the world is warming through carbon dioxide and that humans are the cause.'[13]

The parallels with Ridd's own polemic were remarkable. It was almost as if they were speaking from the same set of notes. As Roberts continued, it became clear why.

We think of the courageous views of scientists who whistleblow on the doomsday scenarios – people such as Professor Peter Ridd of James Cook University. Professor Ridd pointed out that his research questioned the propaganda photographs of the reef paraded by the Centre of Excellence for Coral Reef Studies and the marine park authority, which purported to show long-term collapse of the reef's health, as being potentially misleading and wrong. James Cook University responded earlier this year in an anti-science way by censoring and threatening to fire him. Pre-eminent minds and everyday Aussies – people who speak out – are ridiculed and attacked.[14]

The senator glanced around the chamber, gathered his breath, and went on.

Misrepresentations, exaggerations and personal attacks are natural to the Greens ... What Queenslanders do not like one tiny bit is when one of their own turns on them, like Senator Waters. An August 2016 synthesis by Jim Steele, in one of the most recent peer-reviewed journals *Science*, demonstrated coral reefs can be very resilient, are very resilient, and the gloom and doom claim of Green alarmists is based on unfounded fearmongering. I quote naturalist and essayist, Eric Worrall, who said: 'Given Coral originated 540 million years ago, has survived numerous catastrophic extinction events such as the Permian-Triassic Extinction, which killed around 96 per cent of all marine species, and has effortlessly survived hundreds of millions of years of abrupt natural changes in global temperature, I would suggest the burden of proof—' is on green alarmists [Roberts' words there] – 'to demonstrate why a few degrees' gentle anthropogenic warming is such a threat ... if ... warming actually occurs.'[15]

Roberts was not finished. 'Put simply, coral bleaching cannot occur from human-released carbon dioxide because empirical evidence exists to prove that humans are not causing an increase in global temperatures. The Greens are using control-oriented, elitist media to push their purposeful destruction of the reef's reputation. They are mendacious, destructive and out of order ... What puke,' he concluded.[16]

The Eric Worrall that Senator Roberts had quoted was not, as he'd claimed, the famous Australian naturalist who died in 1984 (whose name is spelled Worrell, with an 'e'), but a self-proclaimed 'software geek' by almost the same name who occasionally posted his views on everything from US immigration policy to renewable energy, climate change and yes, coral reefs, on lower-level climate denialist blogs.* This Eric Worrall had no qualifications in natural sciences, climate change or marine science. The Jim Steele that Roberts quoted was a retired bird scientist from the United States who had no qualifications in coral reef science or any other kind of marine science, and the paper Roberts referred to was not published in the respected journal *Science*, as he claimed, but on Steele's own climate denial website.[17]

Putting Worrall and Steele up against Hughes and other respected scientists was like fielding a team of weekend amateurs against a group of highly talented professional athletes. Pick your sport. You might claim that they will be competitive, but on the field, in a real game, I know who

---

* Worall's X (Twitter) account, @worrall_eric, states that he posts on the climate denialist blog *Watts Up With That?*

I'm putting my money on. Of course, the problem is that the average person doesn't know a Steele from a Hughes, or a Ridd from an Ove Hoegh-Guldberg.

Intentionally or not, Roberts was using two tactics practised by many on that side of the climate war. The first was the *misinformation ambush*. This involves the surprise introduction of obscure, difficult to find and verify, and often inaccurate information presented as fact. Eric Worrall's oversimplistic assertion that because corals are an ancient species and have survived this long, climate change cannot possibly hurt them, is part of a four-paragraph 'essay' posted to an obscure climate denial blog. The ornithologist Jim Steele's so-called synthesis of coral reef science comprised his own views, and was posted to his own website.[18]

The second tactic they used was something I call *credibility camouflage*. To provide your argument with a veneer of credibility, embed a reference to a well-respected, reputable person or publication. It doesn't matter if the subject or conclusions are relevant to your argument, or even support it – or, as in the case of Steele's article allegedly in the journal *Science*, whether the reference you are describing is even real. It's enough simply to place the mention adjacent to, and preferably within the same sentence as, your argument.* Misrepresentation indeed. This is credibility by association, and as I would soon learn first-hand and to my detriment, some people will go to bizarre lengths to achieve it.

This is how truth dies. A little at a time, increasingly obscured by the fog of misinformation.

~

---

* By 2024, the AEF's 'About Us' webpage included mention of some of the principles of the highly respected International Union for the Conservation of Nature (IUCN), while attacking climate science a few clicks away.

Meanwhile, the attack on reef science and science in general was gaining momentum, not just in Australia but around the world.[19] In an op-ed for Fox News in February 2018, Peter Ridd restated his Sky News comments: 'We can no longer trust the scientific organizations like the Australian Institute of Marine Science, even things like the Australian Research Council Centre of Excellence for Coral Reef Studies ... the science coming out is not properly checked, tested or replicated.'[20] He went on to claim, without providing any proof, that 'much of the "science" claiming damage to the reef is either plain wrong or greatly exaggerated'.

Over the next few days, he followed up with a barrage of guest appearances on Sky News and other outlets, blinking behind his glasses, earnest and grey-haired. Clearly, he was not a student of irony – few of his assertions were backed up by even the most basic peer-reviewed science, and hadn't been checked or replicated in any way.

His Fox News piece also took aim at peer review itself, the system of independent checking of work that has been the foundation of modern science for decades, and is in part responsible for the flourishing of scientific progress in all disciplines around the world.[21] 'Nowadays,' he wrote, '[peer review] usually takes the form of a couple of anonymous reviewing scientists having a quick check over the work of a colleague in the field.'[22] Peer review was, in his opinion, suspect and not to be trusted, and therefore the science was also suspect. I wondered if he thus considered his own published papers suspect too.

Again, Ridd seemed to be doing exactly what he accused others of doing – misleading the public. This after all is one of the essential elements in the climate denialist handbook: sow doubt and confusion. And what better way than to cast doubt upon the scientific process *as a whole*? As always with this approach, there is an element of truth somewhere under the layers of bullshit. Any system that relies on human beings is inevitably flawed and imperfect, as we all are. Perhaps that is partly what is driving our present fascination with artificial intelligence – this notion that perhaps a machine can be free of the imperfections that plague us so, and yet which make us human. But over time we learn. We refine and improve.

The notion that most scientists did not take the peer review process seriously certainly went against everything I had seen in my career as an engineer and environmental scientist. I can still remember the pain I felt as a young scientist, receiving a highly critical review or a downright rejection (I'd had a few). But over time, I'd learned to benefit from the criticism, and found that reviewers' comments could be powerful prompts to improve my work. I also knew that whenever I reviewed a paper, I tried to be as thorough, critical and fair as I possibly could. All the scientists and engineers I had worked with over the years had done the same.

But perhaps the most important aspect of the peer review process, which Ridd appeared to completely ignore, is the fact that the vast bulk of the actual peer review only occurs *after* a paper is published in a journal. By disseminating the work to scholars around the world, the work is then read and critiqued, just as Britta Schaffelke and her co-authors had rebutted Ridd's assertions in his 2018 viewpoint paper in *Marine Pollution Bulletin* criticising the peer review process and calling for a formalised quality-control system for environmental science.[23] The very fact that flaws and even occasionally fraud in science are discovered and made public is not a repudiation, but rather a testament to the robustness of this process.

This notion that everyone working on the Great Barrier Reef, other than himself and a few of his ideologically aligned colleagues, were untrustworthy, fundamentally inferior as scientists, and motivated by ideology and the quest for grant money, was a line that he and others supported by the IPA would continue to push and amplify for the next five years.[24] Put plainly, Ridd was saying he was right, and just about everyone else was wrong.

Inside AIMS, we debated how to respond. Whether to give these statements credibility by responding in kind, through the blunt instrument of the media, or to try to somehow remain above this crude gamesmanship. The urge to hit back was strong. After all, contemporary metrics suggested that AIMS and colleagues at JCU and other Australian universities were producing some of the most impactful and highly cited marine and reef science on the planet. We had just kicked off a new feasibility study for a major research effort into reef restoration and adaptation. One of the partners in that effort, the Great Barrier Reef Foundation, a small Queensland-based charity, had shared their plans to raise up to $100 million to support that research.[25] Momentum in the science community supporting a major new investment package for the reef, which might include funding for reef adaptation, was building. We had a lot to be proud of.

Despite this, and after much discussion, we decided not to take the bait, and to instead let our science do the talking. We would keep publishing our work in the very best, highest-quality scientific journals, and talking about the work we were doing.

But it has always taken more than just science to spark the imagination of the public, motivate politicians to act and build support for protecting nature. In the face of those who seek to discredit science, delay action or indeed promote the destruction of our remaining wilderness in the name of economic progress, it is people who make the difference. Individuals dedicated to giving a voice to the voiceless, committed to acting for the benefit of all rather than for a handful of shareholders. And as I was about to learn, the history of the reef is full of people whose vision and drive have had a profound and lasting positive impact.

5

Visions for the Reef

A couple weeks after Ridd's Fox News piece appeared, I was visited by a friend and colleague who had journeyed from his home in country Victoria to spend a few days in northern Queensland. At dinner one night I shared with him some of what had been going on out on the reef and in the media. I confessed that all of these attacks and the deliberate attempts to mislead the public had me rattled. The level of rancour and outrage that seemed to accompany every mention of the reef had surprised me and had left me, if I'm honest, profoundly saddened. I admitted to him that I was worried about the fate of all of the world's coral reefs.

The latest data from our GBR Long-Term Monitoring Program was coming in and the damage from the back-to-back bleaching events of 2016 and 2017 was clear to see. Coral cover had declined significantly in the northern and central parts of the reef. In total we had lost about half of all of the shallow corals on the reef. Sure, we knew from studies we had done on Scott Reef in Western Australia, which had been devastated in the 1998 El Niño mass bleaching and lost almost 90 per cent of its coral cover, that reefs can slowly recover over a decade or so. We also knew from places like Hawaii and the Caribbean that some reefs, once severely damaged, would never recover.

What scared me most, I told him, was that there was still so much we didn't know about these amazing things we called reefs. What makes one coral resilient and another not? What allows one reef to bounce back quickly after being hit by a cyclone or a bleaching event, while others languish and then finally die, leaving behind a skeleton graveyard of grey rubble and choking algae? And what will all of these changes bring – this unprecedented accumulation of heat in the oceans, with the no-change-in-sight acceleration of global emissions driving up temperatures? Can we possibly model such a complex system well enough to provide some predictions? There was so much to be done, I told him, that sometimes I didn't know where to start, or where we could possibly find the money to do everything we needed to do. And if those voices attacking science gained traction with the public, as they already seemed to have done with some politicians, what hope did we have?

I shared with him that inside the scientific community, the discussion had shifted very rapidly in the last couple of years, away from conservation and towards the possibility of actively intervening to help coral reefs adapt to climate change. The idea was to buy time for coral reefs by making them more resistant to rising water temperatures. If we could do that, then maybe they could hold out until the world finally got emissions under control. But the latest predictions from the IPCC suggested that the likelihood of holding global average temperature changes below 2 degrees Celsius was becoming vanishingly small.[1] If we didn't start the research now to see if we could help reefs adapt, then by the time the politicians finally decided that we needed to do something, it would be too late. All we would be able to do was shrug our shoulders. Our scientists figured it would take us ten years of research and development to come up with safe, effective and affordable solutions. And meanwhile, even the most basic facts about what was happening out there were being loudly contested in the media. It was more than frustrating.

My friend is a historian by profession, now semi-retired – a man of deep experience. He sat a moment and stared down into his glass of red wine, swirling the contents. Then he looked up at me and smiled. 'You know that from a historical perspective, none of this is new,' he said. 'All of this conflict.'

I waited for a while, and he continued. 'It may not be much of a comfort, but at least you can know that you are not the first person to have faced this, and that you are not alone.'

'You're right,' I said. 'Not very comforting.' I still could not get over how *personal* it all felt.

'I'll bring a book around for you tomorrow. You'll see what I mean. Written by the first man to set out the idea of setting up a park to protect the reef. 1908.'

'So it only took eighty-odd years,' I said.

'Not bad, idea to reality.'

I shrugged and finished off my whisky. 'I guess. Problem is, the way things are going we don't have another eighty years to sort this out.'

'Perhaps. Here's one for you, just to keep it all in context.' My friend is a great believer in context. 'The American Petroleum Institute, the API – have you heard of them?'

I nodded. 'I did some consulting work for them once.' A long time ago, in another life.

'Well, in 1998, the year of that big global mass bleaching you were talking about, the API issued a document called the *Global Climate Science Communication Action Plan*.[2] In it they described how environmental groups had got control of the climate change science debate, and they called for members – all the big oil companies basically – to target media and politicians with briefings from climate sceptic scientists, focusing on the many uncertainties in the science that still remained. Sound familiar?'

'Jesus,' I said.

He saw the look on my face and didn't push me.

Eventually, I said, 'I attended the World Petroleum Congress conference in Beijing in 1997. I was presenting a paper about remediating groundwater contaminated by gas condensate, co-authored by our oil industry clients. Lee Raymond, CEO of ExxonMobil, was there. I remember being in the audience when he said that there had been no warming since the 1970s – that the earth was cooler than it had been twenty years previously. I was furious. I almost got up in front of fifteen hundred people and told him he was a downright liar.'

'But you didn't.'

I hung my head. Shame flooded through me. 'I was young and most of our clients were oil companies.'

'And by 2001, just a few years later, George Bush, who had originally been a key supporter of global climate action, pulled the United States out the Kyoto climate agreement.'

'Two wasted decades.'

My friend reached out and put his hand on my shoulder. 'Don't,' he said. 'We are all part of this. We all put up with it, every day, make each of those little decisions which add up over the years. The question is, what are we going to do *now*?'

We had a few more drinks and each went our own way. I slept badly, as I always do when I drink and think too much about the future.

The next day, my friend dropped by and handed me an old banged-up paperback copy of Edmund Banfield's *The Confessions of a Beachcomber*, and urged me to read it right away. 'It will help,' he said, 'with everything.'

Born in Liverpool, Banfield immigrated to Australia with his parents in 1856. After moving to Townsville in 1883 to take up journalism, he immediately became interested in the reef and wrote several articles and promotional pamphlets about the splendours of the natural environment of North Queensland. Beset by health problems, Banfield and his new bride started taking short camping breaks on Dunk Island, about two hundred kilometres north of Townsville. Enthralled with the tropical paradise he found there, and encouraged by his wife, he set up a holiday house on Dunk Island, and in 1900 was granted a long-term lease on the property. He quit his job, moved to the island and lived there until his death in 1923, living a simple life of self-sufficiency and reflection.

Confessions, published in 1908, was an immediate success. The story of a Robinson Crusoe–type existence (complete with Tom, an Aboriginal worker who helped the couple survive), sparked the imaginations of European readers. The book describes the couple's back-to-nature existence in a tropical paradise, a turn-of-the-century version of Henry David Thoreau's *Walden*. Full of detailed observations of the island's flora and fauna, and loving descriptions of the reef, the book is both an ode to nature's power to heal and a personal statement of independence from the offices-and-concrete straitjacket of Western society.

Over the next few days, I devoured the book. It seemed to act as a salve. The undercurrent of dread and guilt that had been shadowing me for some time slowly dissipated. Certain passages grabbed me: 'The idea of retiring to an island was not spontaneous. It was evolved from a sentimental regard for the welfare of bird and plant life,' and in rejection of 'the offences which man commits against the laws of Nature'. I couldn't help admiring the courage it must have taken to journey out so far into the unknown, to break with all of the conventions of the times, to risk the disapproval of family and peers, suffer the privations of living completely off-grid and stand up so forthrightly for what he believed in.

I also found a quote from Banfield where he imagined 'a great insular national park ... A park not to be improved by formal walks ... In such a wilderness the generations to come might wander, noting every detail – except in relation to original population – as it was in Cook's day and for centuries before.'[3]

Searching for context, I decided to dig deeper into the past. I scoured the library for stories of the reef. I read whenever I could, snatching a few moments on a flight, at night before turning out the light. Transported back more than a century, I followed the exploits of a small group of people dedicated to doing the right thing in the face of the entrenched views of a deeply conservative society, and with minimal financial support.

~

Banfield, though a keen observer of nature, was not a scientist. His calls for protecting the reef were philosophical and aesthetic, and resonated with a public fascinated with the notion of a still-wild world where mankind was still the underdog. The cataclysm of the First World War came and went, Banfield remaining snug on his little island as the world changed irrevocably around him. He watched as the promise of economic opportunity led to growing exploitation of the reef. Despite its almost unimaginable size and diversity, the signs were emerging, even back then, that human activity was causing severe damage.

Increasingly worried, Banfield corresponded regularly with the small group of Australian scientists working to understand the mysteries of the reef. One of these men was Charles Hedley of the Australian Museum in Sydney. Hedley was born in Yorkshire in 1862, and immigrated to Australia in 1882. A naturalist who specialised in the study of molluscs and coral reefs, he worked as a conchologist and in management at the Australian Museum in Sydney from 1891 to 1920, publishing extensively on the molluscan fauna of Australia and the Pacific islands. In 1896, he participated in the Royal Society's coral reef boring expedition to Funafuti (which is today called Tuvalu). In 1919, increasingly concerned by the dangers of uncontrolled exploitation of the natural bounty of the reef, he wrote a letter to the Queensland Government regarding what he saw as the out-of-control exploitation of the Torres Strait pearl fisheries. In that letter he made an analogy to the deforestation that Queensland was witnessing at the time, where 'at the start only the best specimens are taken', but as pressures and prices rise 'they would if unchecked take every last tree and not replant, until the forest is destroyed'. He concluded that control of the industry was needed to prevent a complete collapse.[4]

In 1920, Hedley and four other Australian scientists attended the first Pan-Pacific Union Congress in Hawaii, whose main objective was to take stock of the present state of knowledge of the biology, geology and anthropology of the Pacific. A sign of the times, the stated purpose was to better support the development of the region's economic potential. There, he presented a paper arguing for the establishment of a permanent island research station on the Great Barrier Reef, and spent time with a fellow antipodean delegate, Australian-born Henry Richards, a lecturer in geology at the University of Queensland. The congress concluded with a statement advocating for increased efforts to understand the ecology of the tropical marine environment and the need for conservation. According to the delegates, 'In the case of the Pacific Ocean, certain marine economic species have been exterminated and others are in peril of extinction or grave depletion.'

Back home, motivated by all that he had heard in Hawaii, including from Hedley, Richards published a paper entitled 'Problems of the Great Barrier Reef'.[5] In it he described the paucity of scientific knowledge of the reef, including its geology, provenance, ecology and economic potential.

He outlined the dangers of unbridled access, stating that 'the exploitation of the Great Barrier Reef has gone on and we stand idly by'. The impact of the paper was far-reaching, eventually leading to the establishment of the Great Barrier Reef Committee of the Royal Geographic Society of Australia, designed to further serious and sustained research on the reef. Richards was named the organisation's first honorary secretary, and in that capacity became the first person to seek government funding from the Australian Council for Science and Industry, the initial precursor of the CSIRO. Hedley became one of the committee's first members.

In April 1923, the new committee laid out its ambitious plans for Australian reef science. A scientific expedition was conceived, the first of its kind, with a focus on understanding the geology of the reef and its economic wealth, namely the already extensively exploited pearl shell, trochus and sea-cucumber fisheries. Pearl shell and trochus shell had been harvested for centuries by Aboriginal people, but in the late 1800s shell became popular in Europe, where it was used to make cutlery, buttons, jewellery and other decorative ornaments. Sea cucumber, also known as *bêche-de-mer*, was a highly sought-after delicacy, and by the end of the 19th century, over a hundred vessels were engaged in supplying curing stations up and down the reef.[6] So widespread was the take of these valuable exports that by 1907, the reef was described as 'fished bare',[7] prompting the Queensland government to place limits on the *bêche-de-mer* harvest.

But funds for the 1923 expedition were tight and the issue of a vessel had yet to be worked out. After much wrangling, two berths were secured on the Commonwealth Lighthouse Service's ship *Karuah*, travelling from Cairns to the Torres Strait. Richards immediately nominated Hedley to accompany him. Over five weeks, the two men investigated eleven coral islands and five continental islands, including the Low Isles.[8]

Their report, published that same year, offered some hypotheses about the geological history of the GBR, and reported on a marked recovery of the pearl shell and trochus harvesting grounds as a result of the respite in collection during the war. They warned, however, that more serious scientific studies were needed. They got a chance to make that message clear at the second Pan-Pacific Congress, held in Melbourne later that year. Richards, by all accounts a diligent promoter, took the opportunity afforded by the presence of so many international experts in Australia, and the resulting media interest. He arranged a post-conference reef cruise for selected delegates, including the governor of Queensland and several American and British scientists. Aboard the government steamer *Relief*, the party left Mackay and headed north for Cairns. Over the next ten days Richards and Hedley guided the group as they visited the Palm Islands, the Whitsundays, and seven reefs.

The trip was extensively and enthusiastically covered by the Brisbane press. Both the *Brisbane Courier* and *Daily Mail* featured full-page spreads about the trip, including spectacular underwater photos of the reef, and a description of the group paying homage to Banfield 'the beachcomber', stopping at Dunk Island to lay wildflowers on his grave.[9] The explosion of public interest that followed made Governor Nathan a passionate advocate of the GBRC's plans.

But all was not well. In November of 1923, tired of his outspoken advocacy and unusual methods, and despite his outstanding scientific record, the board of the Australian Museum accepted Hedley's resignation. It was a case of completely opposing views. On the one hand, a majority of highly conservative and authoritarian board members with little or no scientific training aligned with the exploitative ethic of the day, and on the other, a brilliant and passionate scientist arguing for conservation. Several board members resigned in protest, but the decision stood, plunging the organisation into years of well-publicised turmoil.[10]

Luckily, Richards came to the rescue, hiring Hedley as the GBRC's scientific director a few months later. The two wasted no time, proposing a major expedition to drill a series of boreholes on Michaelmas Cay, with the intention of either confirming or rejecting Darwin's reef origin hypotheses. The technical challenges of such a project were significant, and the costs high – so high that the scope was eventually reduced to just one borehole. It wasn't until April of 1926 that drilling finally got underway, with Hedley in charge.

From the start, things didn't go as planned. The going was slow, and expenses mounted quickly. Rather than encountering hard coral, the drilling crews encountered loose coral sand and mud. The need to continually flush the borehole to keep the drilling rods from getting stuck meant that the hoped-for geological samples and fossils could not be recovered. By early August they had drilled as far as they could go, a depth of 177 metres, without finding the anticipated granite base of the reef. The GBRC's money was gone, with little to show for all the effort.

Other members of the committee thought that too much time and money had been spent on trying to understand the geology of the reef, and not nearly enough on the unique biology of reef-building corals. Now that the GBRC was essentially broke, the rift quickly grew into an all-out row between the geology-focused committee and the host organisation, the Royal Geographical Society of Australia, Queensland. Despite a £1000 bailout from the Commonwealth Government, the disagreement continued to simmer for years, eventually leading to the formal separation of the two bodies.

And then, in September of 1925, the reef lost one of its most powerful advocates, when Nathan ended his term as governor and returned to his native England. A year later, the reef lost arguably its most famous scientist of the age when Charles Hedley died. His ashes were cast across his beloved reef, and meaningful scientific research on the GBR ground to a halt. The next major expedition to the reef would have to wait until 1928, when Maurice and Mattie Yonge and their team arrived on Low Island. Efforts to protect the reef always seem to come in fits and starts, driven by passionate individuals and citizens' organisations, and variously helped and hindered by governments.

6

The Fight for Funding

It was now 2018, and I'd been CEO at AIMS for a little less than two years. Donald Trump was in his second year as the United States' 45th president, and Malcolm Turnbull remained, for now, Australia's prime minister. The mass bleaching on the reef over the past two years was now world news, and everything we were seeing confirmed that the extent of the damage was as bad or worse than we had feared.

As if to underscore the sometimes-ghoulish fascination with this particular kind of bad news, I got a call over the Christmas break from a reporter from the United States wanting to know when would be the best time to come out and see the corals fluorescing as they died. When corals first start to bleach, the expulsion of the symbiont creates a spectacular multi-coloured fluorescent display. The corals literally glow in the dark in a hauntingly beautiful cry for help. When I explained that so far this summer we were not expecting bleaching, and that we could not be sure when this might occur again, he seemed disappointed.

Not long back from Christmas holidays, I was informed by a breathless parliamentary liaison officer that we were to receive a visit from none other than the prime minister himself. Exact dates and itineraries were still to be worked out, but Malcolm Turnbull wanted to visit AIMS to announce a new suite of funding measures for the reef. We hadn't had a prime ministerial visit at the Cape Cleveland site since Bob Hawke famously arrived in a helicopter in the 1980s.

In the end, things moved very quickly. Before the month was out, the PM arrived at our site by car, toured the facility, and at a ceremony in our main library hub, announced funding for a number of new reef measures, including $6 million for AIMS to assess the feasibility of a major research program into reef restoration and adaptation. At the podium, I thanked him and the government, and in a slip of the tongue, referred to him as 'Mister President'. After the laughter had subsided, we took questions from the press, and then it was time for him to leave.

As we were walking back to the main entrance together, I apologised again for my gaffe, blaming my North American heritage.

'It's okay,' he said with a smile. 'I kind of like the sound of it, actually.'

It was only after he was gone that one of my colleagues mentioned that Turnbull was an avid republican and had chaired the Australian Republic Movement for many years.

Through February and March of 2018, we worked hard with the Department of the Environment and Energy and our partners on a proposal to government for major new funding to respond to the events on the reef. It was tough going. Every science agency, university, NGO and reef management organisation seemed to have their own view of how best to respond.

Some argued for more of the same: more effort to improve water quality, building on the decades-long efforts to reduce runoff of sediment, fertiliser and herbicides into the reef lagoon; more boats and divers to attack the crown-of-thorns starfish that literally eat coral; more work with traditional owners of reef sea-country.

Based on results already emerging from the RRAP feasibility study, we and some of the other research organisations were also advocating for investment into new ways to help protect the reef, especially from the threat of climate change. If the 2016 and 2017 bleaching events showed us anything, it was that any benefits of conventional management could be quickly swamped by the damage from sustained marine heatwaves. Climate change had now clearly emerged as the main game, and our current kit bag contained nothing specifically designed to help coral deal with it.

I was spending a lot of time on the phone and travelling back and forth to Brisbane and Canberra, meeting with government bureaucrats and leaders from other research organisations. Slowly, we seemed to be hammering out a package that everyone could live with. But the cost was staggering. When we first presented it, department officials went pale, blinked under the fluorescent lights of sterile meeting rooms, and stammered that, as usual, the government was short of money. The chances of getting anything, we were told, were low. The late-night flight back to Townsville afterwards was a low point.

Over the second quarter of 2018 we continued our efforts to convince government to fund the reef rescue funding package. But as the federal budget approached, everything went quiet. In truth, at this point I had pretty much resigned myself to failure. With each passing day, the odds of our package attracting the funding it needed were dwindling. Despite the efforts of many NGOs and journalists highlighting the need for a major effort to help protect the reef, the news was full of dire statements about the state of the country's finances, and the growing contrarian voices of the climate deniers. It seemed as if we were all going to have to make do, even as we watched one of our most treasured ecosystems battle to recover from such a serious blow.

The question of funding for ecosystem protection is always a difficult one. The onus is on science to show politicians and the public that without help, these places we value are likely to degrade or even be lost altogether. But making a case for *investment* is a lot harder than it seems, for three main reasons.

First, we don't know exactly what is coming next. Predicting the future is a perilous enough business to begin with – when it comes to highly complex ecosystems, it becomes even more difficult. We knew that the GBR had been hit hard, and a lot of coral had died, especially in the northern and central parts. But it was a huge ecosystem, and there was still a lot of work to be done if we were to fully understand the nature and extent of the effects of bleaching. Indeed, our scientists were in the process of designing experiments that could be deployed on the reef the moment we saw the next signs of a serious heat event. So far, 2018 had been a lot more benign out on the reef than 2017 – sea surface temperatures over summer hadn't reached the sustained highs required for mass bleaching. There was still a lot of dead coral out there, but so far, it seemed that the worst was over. But even so, what would the next three years bring, the next ten, twenty?

For answers I spoke with Dr Ken Anthony, a senior scientist and reef modeller based in Brisbane. He and a team of scientists including Professor Peter Mumby of the University of Queensland were working on a series of what he called counterfactuals – model predictions of coral cover on the GBR in the future under different climate scenarios. The work was in its early days, he explained, but the modelling team had made a lot of progress over the past five years. Models had grown in sophistication and in the number of key parameters that they included. Calibration against past datasets, such as the long-term monitoring program records, had shown that the models performed well against what we knew had already happened. And while a lot more needed to be done, these simulations gave us our best picture of what the future might look like.

Ken showed me a series of graphs of predicted coral cover over time, beginning in 1985 at the start of the long-term monitoring record on the GBR, and extending out to the end of this century. The year 2100 seemed like such a long way away. Who could know what the world would look like then? Indeed, neither he nor I would be around to see it. And yet it would come soon enough, as surely as tomorrow's sunrise, as surely as Banfield and Hedley's 1918 had become my 2018.

A series of spiky mountain ranges in different colours jagged their way across Ken's screen. All began at the same point – now – and fell off at different rates as time ticked away to the right. Some of the lines – the ones associated with aggressive global action on reducing greenhouse gas emissions – declined slowly, seesawing at slightly reduced levels of coral cover for a few decades before stabilising. Ken explained that at the moment, with the commitments that governments around the world had pledged so far, we had almost no chance of reaching these levels of emissions reduction. Most of the other coral cover predictions – the ones that described more pessimistic and more likely future emissions scenarios – declined rapidly towards zero, some bottoming out in just the next few decades. It sent a Canadian-winter shiver down the back of my neck.

More work needed to be done to refine these predictions, but Ken was confident that the overall trend and direction of the simulations were robust. It was one of the areas of science that had been included as an investment in our hoped-for package of new reef research funding.

The second reason that obtaining funding for ecosystem rescue is so difficult is that even if we can convince decision-makers that a forest or wetland or coral reef system is in trouble, and that its future without help is a bleak one, we still have to justify the spending. Questions from expenditure committees, finance departments and CFOs queue up in a predictable sequence:

How much do you need?

Okay, that seems like a lot. (Whatever the number is.)

What would happen if we didn't spend the money?

Okay, that sounds bad. But who would care? How would that affect us / our bottom line / our election chances / our social license to operate?

I see. Do you need this right away? What if we just delay this for a few years?

Okay. Then what is absolute minimum we can spend right now, no frills, no extras, just the most critical parts of the program?

And what will we get for that spend? What will we be able to tell our board/electorate/stakeholders?

Sigh. We will see what can be done. But no promises, of course.

The third reason ecosystem protection is difficult to fund is that, quite legitimately, there are a lot of other areas that need investment. People need hospitals and roads and schools and freedom and all of the other things that make a good life in the 21st century. And in this battle for funds, it is often the loudest who win out. Ecosystems and the creatures who inhabit them cannot speak for themselves. Their future depends entirely on those selfless individuals and groups of people who commit to speaking for the voiceless, people like Banfield and Hedley and so many others whose names will never make it into the history books. For all of these reasons, the likelihood of our funding package getting up were looking less and less likely with each passing week.

In April 2018, the weather in Townsville was fine, the summer's heat and humidity finally starting to turn. I was in my office when I received a call from a senior Department of the Environment and Energy official. I confirmed my identity and then listened as the news was explained to me. The official spoke carefully, as if reading from a hastily prepared fact sheet. I took a few deep breaths and asked the official to confirm my plain

English translation of the government-speak. Yes, I was told, the good news was that the government had decided to fund the reef package in full, as developed by the consortium of organisations. Congratulations, you got everything you asked for. It was the single largest investment in the future of the reef and reef science in history. The reef restoration and adaptation research program could go ahead, subject to approval of the business case.

I was bowled over, and for a moment I didn't know how to respond, except to stammer my thanks. After a while I recovered and asked the dangling question: 'So if that's the good news, what's the bad news?'

There was no bad news, I was told. Just unusual news. Rather than the funding for the different major elements of the package going direct to the respective lead organisations over the course of the five years, the funding, all $443 million of it, would go as a single lump sum to a Brisbane-based charity called the Great Barrier Reef Foundation. At the time, the GBRF had been around for almost two decades, and was the largest coral reef charity in the world by revenue.

The funds were to be allocated pretty much as had been set out in our proposal: about $200 million for water quality improvement measures, $58 million for efforts to control crown-of-thorns starfish, $45 million for community and Indigenous engagement projects, and – key for us – $100 million for research into reef restoration and adaptation. Of course, there would be administrative costs that the foundation would charge, but otherwise, all was pretty much as originally envisaged. It would be up to the foundation to administer the funds and distribute them to the various program leads.

I was delighted that the money had been allocated, but I had concerns – primarily over the possibility of delays in getting the money flowing and how the major delivery agencies would be engaged in the process of allocating the funds – and shared these with my board.

The funding decision was announced to the public on 28 April 2018, to almost universal criticism. Over the next several years the foundation would find itself in the eye of a tropical cyclone of controversy, which would damage not only the Turnbull government, but the cause of reef protection more generally.

~

The backlash against the government in the wake of the reef funding announcement was immediate and severe. Everyone had an opinion. The opposition Labor Party and the Greens immediately attacked the process behind the award, claiming that the Turnbull government's decision to give the funds to a twelve-person charity without a tender process was improper and a violation of required government process. They called for the foundation to voluntarily return the grant so a transparent, competitive process could be run.[1]

The debate raged in the newspapers and on social media. In June 2018, a Senate inquiry into the matter was called. The inquiry committee was to be chaired by Greens senator Peter Whish-Wilson, co-chaired by Liberal Jonathon Duniam, and would include members from Labor and the Nationals. Shortly thereafter, we received a freedom of information request from the committee, requiring that we produce all documents and correspondence relating to the funding, including texts, emails and minutes of meetings. Everything I had shared with colleagues about my concerns over the process would be made public. Similar requests were made of the Great Barrier Reef Marine Parks Authority, and the GBRF itself. Public hearings were scheduled for July in Brisbane and September in Canberra.

The foundation's managing director, Anna Marsden, whom I had been working with on the reef restoration and adaptation feasibility study, found herself in the crosshairs. An energetic and intelligent businesswoman, I met her first at the foundation's Brisbane headquarters early in my tenure. Like me, she was relatively new to the reef, having worked previously in a variety of other areas, including the arts. I'd liked her as soon as I met her, and our dealings had been collaborative and constructive. Marsden was quizzed by the media, and then gave evidence at the Senate inquiry in July. She stated that the grant award had come as a 'complete surprise' when first proposed by government, but defended the foundation's ability to deliver the program and to use the grant to raise an additional $300 million for reef projects from private and institutional donors.[2] Our shared view was that we would rather have the money now and be able to start work on all of the urgent projects the reef required than wait perhaps years for another opportunity. It would be up to the GBRF, its staff and all of its partners (including AIMS) to make it work.

The inquiry heard that the decision to provide the funding in this way was communicated to the GBRF at a meeting in April between prime minister Malcolm Turnbull, his minister for the environment and energy, Josh Frydenberg, and the chairman of the GBRF, John Schubert.[3]

In a written statement to the inquiry, the prime minister revealed that the decision to fund the reef through the GBRF was taken so that the government could expense the entire amount in financial year 2017/18, when budget conditions were favourable, rather than over a series of years where budget conditions might not be as good. He also cited the possibility Marsden had mentioned, of leveraging the government's contribution with private sector donations.[4]

Meanwhile, the Australian National Audit Office announced that it would audit the GBRF and determine if appropriate advice was provided by the Department of the Environment and Energy with respect to the award of the new funding. The whole thing was starting to look like a circus with far too many rings.

In September, I was called to Canberra to testify to the inquiry. I arrived the day before and stayed in one of the hotels in the central business district, across the lake from Capital Hill. The weather was fine, so that evening I went for a long run. I headed west along the lake shore and turned upriver towards Scrivener Dam. It had been taking me a good quarter of an hour to get warmed up and into a rhythm of late, but by the time I reached the dam I was feeling good and decided to keep going. Most of my training these days was done at sea level, and at Canberra's 587 metres of elevation the air felt all too thin. It wasn't the years, I told myself, breathing hard, it was the mileage. I kept going. I didn't think about the inquiry, about the death threats and the attacks in the press, about my growing respect for the visionaries of the past. Instead, I let the path spool out under my feet as the sky changed colour through darkening shades of ocean blue into those peculiarly Australian bush pastels and finally into night. By the time I crossed the Commonwealth Avenue Bridge, Parliament House and Capital Hill was lit up like a beacon, and the lights of the city and the passing cars painted the lake.

The next day, David Mead and I arrived early and sat in the back of the room listening to the other witnesses testifying. It didn't take long for climate change to be invoked. The committee heard that some of the GBRF's previous corporate partners had included fossil fuel companies such as AGL, Peabody Energy, BHP and Rio Tinto, whose positions on critical reef issues might not be compatible with the environmental needs of the reef and the foundation's work. A witness from the Australian Conservation Foundation, an environmental NGO, singled out Peabody Energy – a major international coal producer which by that time was no longer a partner of GBRF – pointing out that Peabody had actively funded climate denial groups.[5]

Department officials defended the selection of the GBRF over government agencies such as GBRMPA, CSIRO and AIMS, who could have delivered major parts of the grant. They described the significant monitoring and oversight measures which would be placed on the foundation, and claimed that the Grant Agreement was 'robust and comprehensive'.[6] Several other witnesses disagreed. The Australian Conservation Foundation argued that the GBRF would need to develop the capacity to deliver such a large program, which would take time and money, while existing government organisations could start delivery immediately with existing capacity.[7] The chief executive of the Australian Academy of Science suggested that administering the funds through a government department or agency would provide the required level of transparency and accountability, with very low risk.[8] Other witnesses expressed similar concerns. Why add another layer of bureaucracy and governance, they argued, when the mechanisms to deliver the programs already existed?

Soon it was our turn. I stated my name and position; David did the same. Without much preamble, we were asked if AIMS could have administered the grant if asked.

I adjusted the microphone and leaned forward. 'I'll tell you exactly what I would have said,' I began, too loud. I backed away from the microphone before continuing. 'I would have said: we are probably not the right place to put the whole $443 million because there are a lot of aspects that are operational and we are a research agency. A lot of what's being funded in the other components isn't research. Are we of a scale and do we have the experience and capability to manage, for instance, $100 million for reef restoration and adaptation? Absolutely, we are and we do.'

I reiterated the concerns shared in our freedom of information documents and through our written submission about governance: duplication of effort, delays in starting much-needed research and where the RRAP money would be directed, but restated my confidence in GBRF's ability to get the job done if we all worked together. In the end it was all a bit of an anticlimax. There weren't many more questions for us, and we were dismissed after less than a quarter of an hour. The inquiry wrapped up a couple of days later.

It wasn't until January of 2019 that the Australian National Audit Office released their report into the award. Much to the relief of the government, the auditor-general cleared the PM and Minister Frydenberg of any wrongdoing associated with the $443 million grant. But the ANAO did criticise what it identified as a number of shortcomings in the process, including insufficient scrutiny of the foundation's ability to scale up their activities and the administrative cost of the partnership. They also cited the lack of detail in the agreement, and the failure to involve other partners in the process.[9]

Right around then, it had started to rain in Townsville, hard. And it didn't stop. River levels started rising. Reservoirs filled. Fields and forests, long dry, soaked up the water and became saturated. My son Declan and his girlfriend, Fynn, had chosen this time to visit from Perth. They played tourist for a few days, watching the water rising, getting comprehensively soaked. Soon, water was everywhere. In town, roads and buildings started to flood. The four of us sat in our little house at Cape Cleveland and wondered where all this water could possibly come from. At times the noise on our tin roof was so loud we couldn't hear each other speak. Two days later we were told that the road to the AIMS site was awash, and that levels were rising fast. If we wanted to get out, we needed to leave now, otherwise we would be cut off.

A few staff volunteered to stay behind to keep the Sea Simulator running and keep vital experiments going. In all its fifty-year history, the AIMS site had never been inaccessible longer than three days, so we figured they should be okay. The four of us packed up a few things and left. By the time we reached the causeway, the water was up to the base of the car doors. There was nothing to mark the edges of the road. We took aim for the railway-crossing signs at the highway, and ploughed in. I could feel the current trying to push the car to the right and tried to correct. I tried not to show my fear. Dec and Fynn seemed to be loving it. Eventually we made it to the highway and checked into a hotel in a part of the town that was not yet underwater.

From 29 January to 5 February, in what was described as a one-in-a-thousand-year event, over a metre of rain fell on the Townsville area. The AIMS site was cut off for eleven days. We chartered a helicopter to take in supplies and a fresh crew to relieve the staff who had stayed to keep the site running. I went with them. From the air, the extent of the flooding was clearly visible. Whole suburbs showing only rooftops. Shopping malls inundated, cars floating away in parking lots. The Ross River in full flood, the dam spillway a flying torrent of churning muddy water. I'd never seen anything like it, and a voice inside my head kept reminding me that this was the extreme weather I had been talking about for so many years, that the climatologists had been warning us about for decades: a shudder-inducing illustration of the consequences of anthropogenic climate change.

The Senate inquiry's final report into the $443 million grant followed shortly after it stopped raining in Townsville. The report was accompanied by a set of seven recommendations, and a dissenting report from the Liberal and National members. Sensationally, the main recommendation was that 'all unspent Foundation Partnership funds be returned to the Commonwealth immediately; and that these funds be earmarked for expenditure on projects to protect and preserve the Reef'. If the government chose not to do so, the report recommended that the 'Foundation's investment of public funds precludes investment in sectors or funds that directly or indirectly contribute to climate change, particularly companies that generate energy from or undertake mining of fossil fuels'.[10] And finally, as if acknowledging the proxy war, the committee recommended that the Commonwealth take steps to address and effectively tackle climate change as an underlying cause of economic, social and environmental damage to the reef and the Australian environment more broadly. The dissenting report stated that the senators from the Liberal and National parties did not support the findings of the majority committee report, defending the grant process as suitably robust and transparent.

In the end, if Malcolm Turnbull thought that the announcement of nearly half a billion dollars for the reef would help him politically, he was mistaken. In August of 2018, his leadership of the Liberal Party was challenged, and Scott Morrison emerged as the new leader of the Coalition. After trailing the Australian Labor Party in the polls for months, the Coalition brushed off the growing concern about climate change and its effects on the country, and won a narrow victory in the May 2019 election. Scott Morrison became Australia's thirtieth prime minister. The Morrison government ignored the Senate inquiry's recommendations, and the GBRF began its journey from small charity to major project delivery organisation. But the stains on the government and the foundation would prove hard to wash away.

7

The Threat of Oil

As the controversy over the funding for the reef continued to swirl, I consoled myself with the growing knowledge that these latest events were just one more skirmish in a long battle that had been going on since before I was born. My current reading was Judith Wright's book *The Coral Battleground*,[1] the story of the struggle by a renegade group of activists to save the Great Barrier Reef from oil drilling and limestone mining in the 1960s and 70s.

Wright was born in 1915 in Armidale, New South Wales, to a wealthy pastoral family of British ancestry. Growing up in the country, she developed a deep love of nature and an appreciation of the ways of Indigenous Australians. She studied at the University of Sydney and later worked as a research officer and a literary editor. She published her first book of poetry, *The Moving Image*, in 1946, and went on to produce more than fifty books of poetry, fiction, essays and children's literature. A passionate environmentalist, her writing was characterised by a strong engagement with social issues and a deep connection to the Australian landscape and environment. An activist all her life, she fought for Aboriginal land rights and was involved in various conservation and cultural organisations, such as the Wildlife Preservation Society of

Queensland, the Poets Union and the Australian Council for Aboriginal Reconciliation. She received many awards and honours for her literary and civic contributions, including the Queen's Gold Medal for Poetry in 1991. She died in 2000 in Canberra.

But perhaps what she is remembered for best is her leadership of efforts to protect the Great Barrier Reef. Back in the late 1950s and early 1960s, the GBR was still a vast unexplored and largely misunderstood wilderness. Very little Australian research on the reef had come close to emulating the breadth of insights produced by the Yonge expedition a quarter of a century earlier. But an exploding tourism industry, driven in part by new airline services to North Queensland, was already having negative effects on the reef. In 1947 about five thousand people visited the reef. By 1963, that number had grown to over 125,000. Rampant and unregulated shell collection was all the rage. The reef flats around the Low Isles, where Maurice and Mattie Yonge spent their halcyon year of exploration in 1928, had been stripped bare of coral and shells by tourists who wanted souvenirs. Things got so bad that the Queensland Government finally banned coral and shell removal in 1957.[2]

Channelling Banfield, a small group of concerned volunteers worried about the long-term effects of tourism hatched the idea of a great underwater park that would protect the reef for all time. 'It was an idealistic notion at the time,' said Wright, 'and we were daily pooh-poohed by most people we approached with the suggestion.' At the time, she recalled, the Queensland Government were notionally responsible for the reef, but had provided it with almost no protection. The Commonwealth Government was not interested.[3]

By 1966, Wright's little group had grown, spurred by a general increase in interest in conservation issues that was rippling through Western society. 'It was easy to see that the shibboleth of growth and progress needed a balancing force,' she wrote, 'if the future was going to be lived in a world fit for humans.' Watching the Queensland rainforest being felled and burned, she feared for the reef. The Australian Conservation Foundation was born, with Wright as a founding member. As I read her book, sitting on my balcony overlooking the ocean, I was struck not only by the passion and energy of this amazing woman, but by the parallels with our current situation, more than half a century later. Her words seemed as relevant today as they had fifty years ago.

By 1967 the idea of a marine park in the GBR was gaining steam, but so was the notion that great wealth could be extracted from the reef. Whereas in Yonge's time, the reef yielded wealth primarily through harvesting of its living creatures – trochus and pearl shell, fish and turtles – new opportunities now arose. That year an application was filed with the Queensland Government for removal of coral from Ellison Reef, 35 kilometres north-east of Banfield's Dunk Island, to provide lime for cane farms and cement manufacturing. The proponents had secured advice that the reef was dead, so exploitation would not affect living coral. The ACF and other conservation groups immediately filed objections. 'The fight was on,' Wright remembered. 'It was the first strike in a battle which was to occupy our minds and time for years ahead.'[4] The conservationists promptly raised money for an expedition to Ellison Reef. Contrary to the assertions of the miners, they found a flourishing, living reef, and documented the fact with photos and personal accounts.

But under Queensland mining law the ACF was not classified as an 'interested party', owned no property that might be damaged by mining, and therefore had no legal right to object. About that time, Barry Wain, a young reporter for *The Australian* newspaper, got wind of the story. His reporting focused on the lack of knowledge of the reef, the legitimate threats posed to it and the need for a comprehensive scientific survey of the issue.

Eventually, it all came down to the ruling of Queensland's mining warden. Thanks in part to Wain's reporting, the warden decided to take the ACF's evidence into account, and the application was refused. Ellison Reef was saved.

I put the book down at this point and shook my head. Here we had the same newspaper that fifty years ago helped to save the reef from mining, now providing a megaphone for commentators who claimed that all was well with the reef and that climate change was a non-issue at best, and a hoax at worst. Back then, the proponents of development argued that the reef was dead, so it didn't matter if it was dug up to make fertiliser. Now the assertion was that it was fine, despite all the evidence to the contrary, and so it didn't matter if we kept pumping greenhouse gases into the atmosphere.

Judith Wright and her band of activists, armed with nothing more than passion and a few dollars of donated money, had won the battle of Ellison Reef, but the war was about to move into a new, more dangerous phase.

~

In 1958, Dorothy Hill, a geologist from the University of Queensland, completed a report on the potential for oil deposits beneath the GBR. She recommended that a number of test wells be drilled, and identified prospective locations dotted from Raine Island in the north through to the Capricorn and Bunker group in the south. By 1959 the Queensland Government had introduced generous subsidies to encourage oil exploration in the state, adding to those the Commonwealth had put in place the year before, which meant that taxpayers would pay up to half of drilling costs.

By 1962, Shell, Ampol, Tenneco, Gulf and other companies were already conducting seismic surveys on the reef. And then in 1967, Ronald Camm, the minister for mines in the Queensland Government, quietly zoned all of the Torres Strait and the entire GBR for oil exploration. Without informing the public, the government issued prospecting rights to six companies, with work to start as soon as the following year.

Alerted by the fact that development proponents were approaching the State Government, the ACF got wind of these new permits. Outraged, they started looking into the issue. They soon found that, unlike carbonate mining, oil exploration permits were not open to public submissions. The Queensland Government was adamant – drilling would go ahead. An American geologist, Harry Ladd, was hired by the government to complete a rapid assessment of the reef's oil potential. But before he could complete his report, disaster stuck.

On the morning of 18 March 1967, the captain of the tanker *Torrey Canyon*, en route to Milford Haven in Pembrokeshire in the United Kingdom with a cargo of crude oil from Kuwait, decided to take a shortcut. The ship hit Pollard's Rock – a reef between Land's End and the Isles of Scilly off the coast of Cornwall – ripping open her hull and releasing more than 100,000 tonnes of crude oil into the Atlantic. The slick grew quickly, covering a huge area. When attempts to drag the ship off the rocks failed, the crew were rescued and the UK Government decided to scuttle the ship, burn off the oil and disperse the slick. Over two days, the Royal Air Force spectacularly hit the ship with over a hundred tonnes of bombs, rockets and napalm. Over 2 million gallons of chemical dispersants were poured into the ocean around and on the slick. Twelve days after running aground, the ship broke up and sank, but the damage was done.

The slick eventually fouled much of the Cornwall coast, and came ashore in Guernsey and Brittany in France. The dispersant used by the response teams turned out to be highly toxic to marine life, killing anything it came into contact with. As a result, 85 per cent of France's population of puffins died.[5] The effects of the spill were felt for decades afterwards, and blobs of degraded oil can still be found on Cornish beaches – I found a few myself walking there in 1999. To this day, it remains the United Kingdom's worst environmental disaster.

I can still remember watching it on TV as a small boy: the grainy black-and-white images of hundreds of oil-covered seabirds flapping and stumbling in their death throes, the pillar of smoke from the bombardment rising high into the air, the white-eyed despair in the faces of oil-soaked volunteers wading knee-deep in black sludge as they scooped the oil by hand into pails. Nightmares haunted me for weeks after, and perhaps like many of my generation, helped spark a lifelong concern for the environment.

For the government and the oil companies seeking to find oil under the GBR, the *Torrey Canyon* disaster could not have come at a worse time. The images beamed into homes across the country raised awareness of the fragility of the ocean in a way protests and petitions could not. And for those involved in the burgeoning reef tourism industry, the message was chillingly clear: the environment *is* the attraction. Ruin that, and you have nothing.

Ladd's report was issued in September 1968. His central theme was that the reef's resources, specifically oil and gas, could be exploited while protecting the reef from what he called 'serious and widespread damage'. But, he added, before such development could take place, a comprehensive survey of the GBR was required to evaluate the risk, and to identify actions needed to protect the reef.[6] The Queensland Government's policy of controlled exploitation of oil and gas on the reef was born. And in a weird precursor to the present, Ladd's report identified expanding tourism, coral and shell collecting, and unrestricted fishing as the main threats to the future of the reef, while downplaying those from petroleum production.

But the notion that the oil industry and the government could escape scrutiny was about to be dealt another blow. In October 1968, with a federal election just over a year away, John Büsst and Judith Wright, on behalf of the ACF, called on the Commonwealth Government to take control of the Great Barrier Reef and place a moratorium on all mining and oil exploration for at least five years. They also proposed that a national committee be set up on the future of the GBR, to investigate its potential to 'become the most important marine biological lab on earth, and the most important scenic and recreational area in the world'.[7] The battle was about to get political.

Despite the rising clamour from all sides, the Queensland Government moved ahead with its plans. The first location for drilling was chosen, in Repulse Bay off Mackay. The joint Ampol-Japex well was to spud in beginning October 1969.

And then, on 28 January 1969, just off the coast of Santa Barbara in California, well A-21, being drilled from Union Oil's offshore platform A, suffered a massive blowout. Only the top 239 feet of the well had been cased with steel pipe, rather than the prescribed 300 feet. When the drill cut through into the oil-bearing formation, the tremendous pressure could not be controlled. Crude oil started jetting from the well, and without a steel casing to contain it, high-pressure oil from the reservoir started tearing through the softer layers of rock under the Santa Barbara Channel. Soon, crude oil was pouring into the ocean through a number of ruptures in the sea floor. Within 24 hours, a slick covering 200 square kilometres was visible from the air.

Over the next ten days over 80,000 barrels of oil poured into California's coastal waters, fouling miles of coastline and killing thousands of sea birds, dolphins, seals and sea lions. The spill was front-page news across the US, and received prominent TV coverage worldwide, including spectacular aerial views of the oil slick. The American public was outraged by what stands today as the third-largest oil spill in US history, after the 2010 Deepwater Horizon and 1989 Exxon Valdez disasters.[8]

People in Australia, too, were shocked. If something like this could happen in a place like California, where some of the tightest environmental controls on the planet were already in place, it could happen here. To many, the assurances of government and industry seemed increasingly empty. The Queensland state election was only a few months away, and the government's policy of 'controlled exploitation' of oil on the GBR was increasingly under attack. In March, the opposition dropped another bombshell: the premier, Joh Bjelke-Petersen, and a number of cabinet ministers had direct financial interests in the companies proposing to explore for oil under the reef.[9]

Two weeks before the Queensland election, a major public symposium on the future of the reef was convened at the University of Sydney, organised by the ACF. Over five hundred people attended, including experts in mining, fisheries, geology, tourism and biology, along with the major conservation groups. By all accounts it was a partisan and sometimes fiery affair. Speaker after speaker stood up to praise the reef's unique value, its unquestionable magnificence. But there the agreement ended.

The acknowledged high point of the day was a speech by distinguished Australian jurist Sir Percy Spender, an ex-president of the International Court of Justice and former Australian representative to the United Nations. Spender concluded that Queensland's territorial boundaries ended at the low-water mark, and that under international law the Commonwealth had full legal jurisdiction over the seabed and the natural resources that lay beneath.[10] Queensland mines minister Camm disagreed, making the State Government's position clear: 'We in Queensland claim the Great Barrier Reef and believe we own it.'[11] The delegates might not have known it at the time, but this fundamental legal disagreement over who 'owned' the Great Barrier Reef would become one of the major barriers to progress over the coming years.

But it was not just government that was at odds. All was not well within the conservation movement, and the symposium exposed ruptures that had been growing for some time. As it had in Hedley's time, the schism between the geologists and the ecologists threatened to throw the campaign to save the reef into disarray. The Queensland Littoral Society and the Wildlife Preservation Society of Queensland accused the ACF and the Great Barrier Reef Committee of being soft on the Queensland Government, and John Büsst went so far as stating publicly that the ACF were apologists for the oil companies. Formal protests were filed and eventually the ACF was forced to make concessions. The final recommendations of the symposium included a call for a moratorium on drilling and the establishment of a commission to determine if oil exploration should be considered in the future.

In May 1969, the coalition Liberal–Country Party government was returned in Queensland under Premier Bjelke-Petersen. The planned exploration well in Repulse Bay would go ahead, despite the public outcry.

But by this point, the fate of the Great Barrier Reef was no longer Queensland's to decide, even if its government claimed that legal right. The millions of Australians who had by then visited the reef, and the millions more who had read about or seen pictures of its splendour, now increasingly felt that this was *their* reef as well. Buoyed by Spender's legal view, conservationists saw an opportunity to appeal directly to the Commonwealth Government. In July, John Büsst met Prime Minister John Gorton while on holiday on Dunk Island, and presented the conservationists' case, namely that the Commonwealth should immediately take control of the GBR and place a moratorium on all mining and oil exploration for at least five years.

Emboldened by their success at the polls, the Queensland Government doubled down on its plans. Mines minister Camm stated in the House of Representatives that it would 'not repudiate any agreement that it has entered into', and that 'we can control drilling so that the risk is remote, and it is my intention to do this'.[12]

But to many around the country, recent events in Cornwall and California showed clearly that the risks were real and imminent. If drilling and oil production were allowed on the reef, a major spill wasn't a matter of *if*, but *when*. Conservationists, again aided by the media, including once again *The Australian*, started ramping up the pressure. With a federal election looming, opposition leader Gough Whitlam announced that a Labor government, if successful, would immediately end all mining and oil exploration on the reef. Gorton countered by announcing that if returned, his government would not only end oil exploration on the reef, it would fund an institute of marine science to be based in Townsville (in the marginal seat of Herbert), dedicated to understanding the physical and biological resources beneath Australia's oceans, with study of the GBR as its first priority.[13]

John Gorton's coalition won the October 1969 federal election and retained the seat of Herbert. Almost immediately, Gorton decided to legislate Commonwealth control over the GBR, as proposed by Spender at the ACF symposium in May. States' rights advocates, including in Gorton's own cabinet, were furious. Debate on the issue was scheduled for later in 1970, leading to a High Court challenge of the *Petroleum (Submerged Lands) Act 1967*, one of the key legal planks of the Commonwealth's claims. The Queensland Government remained unmoved. With Gorton's proposed changed still pending, drilling would go ahead as planned.

By January 1970, the Japex drilling rig *Navigator* was approaching the reef, and conservationists dug in for a last-ditch battle. Once again, it was John Büsst who led the way, aided by Senator George Georges. Building on earlier efforts with the Amalgamated Engineering Union, who fully supported the notion of protecting the reef for all Australians, they got the trade unions to place black bans on the *Navigator*. All ports and wharves would be closed to the rig. It would receive no services or supplies of any kind while in Australia.

Once again it was *The Australian* that provided the hammer blow. In an editorial on 7 January 1970, it published Georges's black ban letter to Ampol and the owners of the *Navigator*. It pointed out that the intransigance of the Queensland Government and its unwillingness to heed public opinion would surely backfire. 'The black ban,' it stated, 'will have an unprecedented measure of public support and will probably succeed. It deserves to.'[14]

Ampol had no choice. They backed down. Drilling would be postponed, and the rig was sent home.

Shortly after, as the tension mounted, Gorton agreed to meet with the Queensland premier to find a resolution. Bjelke-Petersen remained defiant, claiming that halting the drilling was unlawful and unnecessary, characterising the public's concern for the reef as 'wild'.[15] Eventually the two men agreed that the only way through the impasse was to convene an impartial inquiry into the affair.

And then, a few weeks later, almost as if pre-ordained, disaster struck again. On 3 March 1970, the Ampol tanker *Oceanic Grandeur,* bound for Brisbane with a full cargo of crude oil, hit an uncharted rock near Thursday Island off the tip of the Cape York Peninsula. Over a thousand tonnes of oil spewed from a 57-metre gash in the stricken ship's hull, creating a slick over 10 kilometres long. The event everyone had feared since oil exploration was first considered had happened: a major oil spill on the Great Barrier Reef.

It took three disasters – one on the reef itself – and a huge effort from a band of self-funded citizens, but in the end, oil exploration and mining were banned on the GBR. The events of 1969 and early 1970 were a tipping point for the reef. Conservationists had won out, for now, in just one battle of what was to be a very long war.

8

Biological and Geological Time

It was early September 2019. The cleanup after the floods in Townsville continued, the cost running into the tens of millions. Controversy over the Turnbull government's $443 million grant to the GBRF raged on unabated. The Australian and Queensland Government's just-released annual Reef Report Card for 2017/18 had found that, despite some areas of improvement, overall water quality in the inshore reef was poor, and progress towards water quality targets was not what it should be.[1] The GBRMPA's 2019 Outlook Report, produced every five years to give a comprehensive look ahead for the GBR, identified climate change and deteriorating water quality from land-based runoff as the biggest threats to the reef's future. The report concluded that without urgent additional local, national and global action, the overall outlook for the Great Barrier Reef's ecosystem would remain very poor.[2]

Just a few weeks earlier, AIMS had issued the 2019 instalment of our long-term monitoring report – the longest-running and most comprehensive accounting of the health of a coral reef system anywhere in the world. The dataset stretched back to 1983, and covered ninety-three mid-shelf and offshore reefs. Thirty-two inshore reefs were also monitored in a separate program. It is work that keeps AIMS scientists like Mike Emslie at sea for over a hundred days a year, every year. He and his team spend countless hours underwater every year, visiting the same reefs and the same locations to ensure consistency, observing the health of the coral and reporting it in minute detail. They are as experienced and knowledgeable about the reef as anyone alive. No sooner have they completed surveying the reef and writing up the year's report than they are planning to get back out on the water to start the process all over again.

Mike was standing in my office looking ill at ease. Tall and lanky, with a face deeply tanned from sea and sun over many years working on and under the ocean, Mike was more comfortable with the rigours of ship life than in an office. Now that he led the monitoring effort and had to front the media on a regular basis, he had replaced his dreadlocks with a close stubble. He wore a tidy AIMS-branded collared shirt rather than an old ragged tee. He was wondering why I'd asked him to meet with me. I explained that we had been asked to brief a backbench committee on Capital Hill. They wanted to hear about the state of the reef first-hand. Who better, I said, to deliver it?

He hesitated a moment and then said, 'We can make up some slides for you. No problem.'

I shook my head. 'I think you should go.'

It was as if I had asked him to run the 16 kilometres from my office along the causeway back to the highway in the oppressive summer humidity, carrying me on his back. It took a while, but finally he agreed.

And then, just as he was about to leave, he stopped himself and asked me what he should wear.

'Well,' I said. 'I reckon you have two options. One, you could show up in your work gear, steel toe-caps and all, as if you've just come from the boat.' He frowned at this. 'Or you could do what I do, and wear a suit and tie.'

He thought about it a moment and shook his head. 'No, not work clothes.'

'No,' I said. 'Probably not.'

'Suit then.'

'Yep.'

'Only thing is, I don't have a suit.'

'Oh,' I said.

'Or a tie.'

In the end, Mike bought a new suit, and a tie to go with it, and delivered his talk. I heard back quickly that he had done an excellent job.

The 2019 long-term monitoring report was titled *Mixed Bill of Health for the Great Barrier Reef*. It described the collective impact of cyclones, bleaching and crown-of-thorns starfish over the last five years, resulting in a reef which was badly damaged, but starting to show the first signs of recovery. The central reefs, from Cooktown to the Whitsundays, had fared the worst. Coral cover, which hit a low of 14 per cent in the 2018 report, a direct result of the extensive bleaching and die-off in 2017, had continued to decline, and was now near a record low at 12 per cent. In comparison, the historical low since records began in 1985 for this part of the reef was in 2012 after Severe Tropical Cyclone Yasi passed through, physically smashing up vast areas of reef. Coral recovered quickly after Yasi, and by 2016 cover was back up to a record high of 22 per cent. The reef is a living organism, and can heal itself and regrow if given the chance.

Coral cover is a measure of the overall health of a coral reef, but it is not the only one. It describes the proportion of the reef area covered by live hard coral. Extremely healthy, thriving reefs can have coral cover ratios as high as 80 per cent. Typically, anything over about 30 per cent is considered good, or in the case of the central part of the reef, anything over about 20 per cent.

The good news was that the northern part of the reef, hardest hit in the 2016 and 2017 bleaching events, was showing the first signs of recovery. Hard coral cover had increased from a worrying record low of 11 per cent in 2018 to 14 per cent in 2019. Not a big increase, but enough to give hope that despite the recent devastation, the reef was still resilient enough to bounce back. Beyond that overall picture, though, concern remained. Some individual reefs, wiped out almost entirely by the 2016–17 events, remained desolate fields of dead rubble. Some might never recover.

The southern part of the reef, which generally enjoyed slightly cooler overall sea temperatures, had escaped the recent bleaching, but was still feeling the effects of Cyclone Hamish in 2009, which caused widespread damage. Coral cover had declined slightly from 25 per cent in 2018 to 24 per cent in 2019 (about half of the 1988 record high of 43 per cent), partly because of an outbreak of the dreaded crown-of-thorns starfish.

It was a nuanced story for a complex ecosystem. There were some signs of recovery which showed the reef was still able to bounce back, given time and the right conditions, but the damage had been severe. The warnings were there for all to see: look after it, curb the emissions that are driving climate change, and try to improve the reef's overall resilience by doing what we can to improve water quality and control the coral-eating starfish, or else we could lose it. The models Ken Anthony and his team were working on were showing this conclusively.

And yet, media commentators on all sides were whipping themselves and each other into a rage. Either the reef was dying, tottering at the very brink, or the reef was absolutely fine, thank you very much. The tone was strident, often angry. All nuance was lost, and each side in the argument focused on the parts of the report that supported its view. In my opening statement to Senate estimates shortly after, I said as much. Few seemed interested in the facts. Scoring points was what mattered. Everyone wanted someone to blame, except themselves. What we needed were solutions – action.

After a while, it all got to be too much. Like many of my colleagues, I tried to tune most of it out and focus on my work. For the last few months, we had been working feverishly with a consortium of research organisations across the country to finalise the business case for investment into reef restoration and adaptation, which was due at the end of the year. Key to the report were the counterfactuals Ken and his team were working on – model outputs that tried to predict what would happen if we did nothing while temperatures keep rising. The existing models, developed by the University of Queensland and CSIRO, were relatively crude, but at that moment they were the best we had. And what they were telling us was frightening. Depending on the emission pathways the world follows over the next several decades, the GBR could be reduced to a mere shadow of its current already-degraded self by mid-century. The more greenhouse gases we pumped into the atmosphere, the worse it would get. In many of the scenarios, the reef was essentially gone by the end of the century. But scenarios that modelled basic interventions of the kind we were proposing showed that we could buy years and even decades of time for the reef, while we worked to drive global emissions down. Our work showed clearly that intervening on the reef at large scale was technically possible, but not just yet. We simply didn't know enough about the methods themselves, how they might work, what the costs and risks were. To get to that point would take at least five years of focused research and development, and money – lots of it.

At this point, I had taken over drafting the all-important business case document, which would determine whether Australia invested as much as $100 million from the Turnbull government's grant to the GBRF into helping reefs adapt to climate change. David Mead was bringing together all of the technical information and evaluating the various interventions we were looking at. There were over forty intervention possibilities – everything from cloud brightening (pumping seawater vapour into the the air to create low-level clouds that will reflect light during marine heatwaves) to selectively breeding heat-tolerant corals to deploy on the reef.

But we had hit a major hurdle. Professor Ian Chubb, chairman of the government's independent expert panel on the reef, was insisting that the report be externally peer-reviewed by experts who had no vested scientific interest in the reef – not 'reef people', as he put it to me in a meeting in Canberra. We had identified a list of eminent overseas scientists from other disciplines – hydrology, genomics, engineering – to approach for help.

Long days blended one into the other, and I spent late nights in front of the computer screen, leaving me bleary-eyed and ragged. The deadline for the report was approaching fast, and we were running out of time. One night I came home late and collapsed onto the couch. I had planned on going for a run that night to blow out the stress and the frustration, but it was too late now, too dark out here on our lonely peninsula, and I was exhausted.

Heidi came and sat next to me, handing me a beer. 'You look like you need a holiday,' she said.

'Too busy,' I mumbled, not feeling like the best company.

'Two years without a break,' she said.

'We were in Canada just last Christmas.' I was being defensive now.

She nodded, sipped her wine. 'Let's go somewhere different. Do some hiking, swim in the sea. Just us.'

'I've always wanted to go to Lord Howe Island,' I said. 'It has the world's most southerly coral reefs. What do you think?'

The next day I booked a week on Lord Howe for October of the coming year. In the end, it would take us almost three years to finally get there.

~

With the Lord Howe trip still a way off, when the chance came to get out of the office and away from the firestorm of inquiries and social media attacks, I jumped at it. The constant bickering and posturing was getting to me. The more political the issue became, the less the truth seemed to matter. It is sometimes said that truth is the first casualty of war, and I was about to learn that this was a war that had been going on for as long as Europeans have been in Australia.

When I visited the Keppel Islands as part of an AIMS scientific expedition – the same islands visited by that group of One Nation senators in 2016 – I found a place that looked pretty much as it had back in Yonge's time. Azure waters fringed the white beaches and green hills of the two largest inhabited islands, Great Keppel (Wop-pa) and North Keppel (Konomie). Smaller uninhabited islands dotted the surrounding waters, their rocky crags glowing in the late afternoon sun. Fish and turtles drifted peacefully among coral gardens that merged into the basalt cliffs rising from the water. We found the reefs that fringe the islands in excellent condition, as the senators had, despite their proximity to the mouth of the Fitzroy River and the area's popularity with fishers.

These reefs had escaped the worst effects of the mass bleaching events of 2016 and 2017, which had hit the northern and central GBR hard. This alone was a powerful reason to come here and study what made these corals particularly hardy and resilient.

This group of eighteen islands and surrounding reefs is a popular recreation and holiday destination. The southernmost island of the group, Peak Island, is a major nesting site for flatback turtles, and hosts one of the two largest nesting populations that remain in eastern Australia.[3] If you are lucky enough to visit Wop-pa, you can not only enjoy the remarkably resilient reef that sits just off the beach, but you can visit a permanent museum of Woppaburra history and culture.

Bob Muir, a Woppaburra elder, generously showed me around his country. Quietly spoken, with a warm smile that beams out through a thick greying beard, he talked openly about his people's diaspora and the challenges they have faced for recognition. Scattered across the country, reduced almost to extinction, it took decades for the Woppaburra to rebuild as a people. As he talked about his people's journey, the catch in his voice betrayed the emotion. Much was lost along the way, he told me. The old stories died, gone with the last of the elders. With each passing year, their sense of place and belonging, so important to all human beings, erodes away. The museum is the Woppaburra people's way of holding on to what they have left.

As I walked around the little building with Bob Muir, I could feel the original inhabitants of the place staring out at me from the black-and-white photographs on the walls. I could hear their voices echoing from the artefacts in the cabinets, sense those centuries of history in the warm trade winds blowing across the islands. But it was the totem just outside the museum that affected me most. It was a simple thing, a native log perhaps four metres high, planted in the ground just outside the front door. Painted all in black, standing in stark contrast against the azure sky and the opaline waters of the reef, it exuded a remarkable and sombre power. I stood looking at it a long time.

Later that day, just as the boat was arriving to take us back to the mainland, I asked Bob what it meant. He looked at me a moment, then let go that smile of his. 'Have a look at the top,' was all he said, never a man of many words. 'You have time.'

I ran back to the museum. And there it was. How could I have missed it? Right at the very top of the black pole, spanning time and history, was a thin band of white paint, perhaps no more than a centimetre in width – the blip of white settlement in the vast black history. I took a couple of steps back, breathed deep, nodded to myself a few times, and walked slowly back to the beach. By the time I got back, the boat was loaded and ready to go. The others were waiting for me. I could see Bob looking out at me from the bow. One glance was enough. He knew that I understood.

We need to change our relationship with this past, not by ignoring it or leaving it behind, but by somehow creating a new partnership that respects the vast history that has come before.

It wasn't until 1993 that the Woppaburra had rebuilt sufficiently as a people to take their first steps towards reclaiming the future. That year, a few of the descendants of those evicted in 1911, led by a young Bob Muir, returned to Wop-pa and solemnly raised the Aboriginal flag on the beach, symbolically reclaiming the island's unallocated State land for the Woppaburra People. It took another fourteen years for the Queensland Government to officially hand back 174 hectares on Wop-pa, along with $250, to the traditional owners.[4]

It wasn't much – just a small piece of their traditional lands – but it was a start. They formed the Woppaburra Land Trust to manage their new-old patrimony. They took the first tentative, largely self-funded steps to repairing some of the damage done by decades of neglect and misuse. They approached the Queensland Government for help and launched a project to protect a culturally important midden on the island, which was being destroyed by four-wheel drive vehicles taking tourists snorkelling on the reef off Monkey Beach. Fencing and bollards helped divert much of the traffic, and were then eventually replaced with boardwalks, permanent fencing, and interpretive signage. It doesn't sound like much, but for the Woppaburra, it was physical evidence of progress.

Since then, the Woppaburra have come a long way. They have signed a Traditional Use of Marine Resources Agreement with the Great Barrier Reef Marine Park Authority and the Queensland Government. They worked with GBRMPA to develop guidelines that help the marine park authority include cultural heritage in their permitting and processes, and give direction to other organisations seeking permission to conduct research or other projects on Woppaburra country. And in 2021, they would finally be granted Native Title by the Commonwealth over much of their island country.

Yonge and his expedition came to the GBR all those years ago, the first European scientists to study the reef. But Bob Muir's people have lived here for millennia, in harmony with the reef and its creatures, since long before Europeans even knew *Terra Australis Incognita* existed.

~

In the main entrance of the AIMS headquarters building, not far from my office, there was a seasonal calendar developed by the Masigalgal people of Masig Island in the Torres Strait. Their elders recognise four distinct seasons, Naigai, Zei, Kuki, and Woerr, which vary in timing and duration year by year. The end of one season and the start of the next depend on observed changes in weather, winds, and the appearance and abundance of plants and animals. Survival depended not only on understanding these changes, which dictated what could be harvested and when, but on observing the fundamental philosophy of *Gud Pasin* and *Mina Pawa* (good ways and appropriate behaviours) that lies at the core of the Masigalgal system, passed down as ancient knowledge from elders, generation after generation.

The Masigalgal calendar is in the shape of a wheel that turns from one season to the next. Radiating from the centre is a series of beautiful hand-drawn paintings. The outermost ring depicts the changes to the island's plants. The next ring describes the chief activities of the people, then the behaviour of the island's birds, and finally, near the centre, the changes in its surrounding seas. In late August and early September, for instance, as Naigai moves into Woerr and the Cedar Bay cherry begins to flower, the traditional time for house building and maintenance has come. Soon, sooty oystercatchers will fly northward, and seeing this, the islanders know that hammerhead sharks will appear on the surrounding reefs. As the days go by, and the Kubil Gim fruit ripens, turtles will start mating, and native yams can be harvested. And so it goes, an intimate rhythm that links the people to the island as they move together through time.

And yet this calendar is more than a beautiful work of art, more than a practical guide to living and prospering in one of the most remote places on the planet. The Woppaburra and the other sea-country people of Australia depended on the ocean for survival. Over millennia, they developed a sophisticated understanding of its rhythms and relationships. They knew the best places and times to harvest fish, and how much to take to avoid depleting those stocks. There is ample evidence that Aboriginal peoples built complex fish traps, farmed crops and compiled deep observational knowledge of the world around them.[5] This knowledge of how nature worked, passed down faithfully over generations, was underpinned by a sophisticated philosophy of responsibility, co-dependence and caring for country.

But by the time Maurice and Mattie Yonge arrived in Australia in 1927, much of that ancient knowledge had been lost, was in the process of being lost, or was being actively ignored and in some cases even suppressed. Europeans brought a new ethic to this ancient land when they arrived in the late 18th century – one based on commerce, expansion and the unregulated exploitation of the natural world.

Early European explorers in Australia described bays and inlets where there were so many fish you could literally wade into the water and scoop them up with your hands. Captain Cook's diaries are replete with descriptions of a stunning abundance of fish along Australia's east coast. He writes of regularly catching fish weighing 200 pounds or more, sharks and rays bigger than any he had ever seen, and vast heaps of giant oyster shells harvested by the natives. The oceans were a cornucopia of undiminishable wealth, there for the taking.

And so we took. And the more we took, the more the oceans gave. The nimiety was staggering. The notion that we, small and puny with our clever brains and simple machines, might ever reduce or in any way diminish these mighty God-given infinities seemed pure lunacy. Even today there are those that still claim, with twisted humility, that there is simply no way that mere mortals could ever affect anything as big as the atmosphere or the oceans.

A century or more ago, that view might have been quite understandable. After all, the oceans cover 71 per cent of the surface of the earth – 360 million square kilometres of sea, much of it cold, lonely and windswept. That's over a *billion* cubic *kilometres* of salt water. The oceans are, on average, about 3.5 kilometres deep, and the deepest trenches plunge to bone-crushing depths of up to 11 kilometres. Even today, as much as 80 per cent of our oceans remain unmapped and unknown. So vast are they, that we still have not come close to knowing how many species live there. In a decade-long global census of marine life completed in 2010, scientists identified over 1800 previously unknown species. Currently, over 230,000 marine species have been catalogued and accepted by the World Register of Marine Species.[6] Scientists estimate that as many as two million more species remain unknown. The world's oceans today feed over three billion people.[7] Over half of all the oxygen produced on earth comes from the ocean, the majority from oceanic plankton. One species, *Prochlorococcus*, the smallest photosynthetic organism on earth, alone produces as much as 20 per cent of the world's oxygen – more than all tropical rainforests combined.*

The oceans provide the basis for life on earth. They interact with our atmosphere to regulate our climate and weather, and provide the fundamental building blocks of the global food chain. For all recorded history they have provided us with a seemingly inexhaustible source of food and wealth. And there is that other thing, of course. Since the dawn of humanity, the oceans have fired our imaginations, challenged and terrified us, and provided a sense of inspiration and meaning to our lonely existence, here on the only known inhabited planet in all of the universe.

But we don't live a hundred years ago, and we must stop thinking as if we do. The climate deniers like to use the geological record to show that climate change is natural, and has been going on since our planet's earliest beginnings. And in that sense, they are right, of course. But those changes, driven by cycles of solar radiation, volcanism and the slow evolution of

---

* I got some of the facts for this section from the United States' National Oceanographic and Atmospheric Administration's (NOAA) Ocean Service website. In the frequently asked questions section of the website, I came across the question, 'Is the Earth round?'

our planet, occur over hundreds of thousands or millions of years. They occur in geological time.

But we don't live in geological time. The whole of human history since our emergence as a species is barely a blip on the geological clock. We exist in *biological* time, measured not in epochs or eons, but in the short spans of generations, a few decades for human beings, mere months or days for many of the other creatures we share the planet with. And the changes to our atmosphere that are driving climate change now are occurring in biological time, with profound effects here and now.

Each of us, in turn, must learn to live within the time span life has provided us. And, crucially, we must learn to live within the means our planet has provided us.

9

Tragedies of the Commons

In Herman Melville's classic 1851 novel, Captain Ahab is obsessed with finding and killing the 'great white whale', whom he calls Moby Dick, the largest sperm whale that ever existed. Published at the peak of American whaling and regarded as a classic of American literature, *Moby-Dick* is a novel about the follies of man, including the ruthless exploitation of the natural world.

The history of whaling in Australia is both an analogue for the broader exploitation of the oceans, and a salutary example of what can be achieved through conservation. Incredibly valuable, and with no rules limiting the catch, whales were systematically hunted to the edge of extinction, place by place, species by species. It was, quite literally, a free-for-all. Under such conditions, any resource is doomed. It was a classic example of a phenomenon known as the tragedy of the commons.

Popularised by economist Garrett Hardin in 1968, the concept of the tragedy of the commons goes back to the time of the ancient Greeks.[1] It states that when there is free, unrestricted access to and demand for a finite resource (such as whales or fish, or even the capacity of the oceans to assimilate pollution and heat), then it is structurally doomed to collapse through over-exploitation.[2]

By the time Yonge's expedition arrived in North Queensland in 1928, commercial whaling in Australia had evolved considerably from its 18th-century sail and manpower origins. The development of harpoon guns, explosive harpoons and steam-driven whaling boats in the 19th century made large-scale commercial whaling not only safer for the men involved, but much more efficient and profitable.

Back then, the centre of Australia's commercial whaling industry was in Albany, on the southern coast of Western Australia. Built on a hill overlooking the magnificent expanse of King George Sound, Albany was the first major settlement in WA, pre-dating both Perth and Fremantle, and for a long time it was the state's only deep-water port. Founded in 1826, the town's main street rises up from the sound, lined with the beautifully restored facades of pioneer churches, homes and public buildings. Anzac troops marched down this street on their way to embark on their fateful journey to Gallipoli in the First World War. And from 1912 until 1978, one of the town's major sources of employment was whaling.

In 2010, I visited the whaling station at Albany with my family. Converted into a museum years ago, the place gives you a sense not only of the size of the whales themselves, but also of the highly mechanised industrial nature of the slaughter that occurred there. One of the things that strikes you first is the smell. Even decades after the last whale was flensed of its blubber and the carcass dumped into the sound for the sharks, the uniquely oppressive odour of whale blubber hangs on here, seeping from the cracks in the concrete flensing ramps and the fibre of the wooden beams in the boiling rooms, where the fat was boiled in huge cauldrons.

On that same trip, we were lucky to spend a full hour in King George Sound within touching distance of three mature southern right whales as they floated serenely under a grey overcast sky. Peering at us through big, dark, heavily lidded eyes, they seemed as interested in us as we were in them. For their tremendous size – adult females can reach up to 18 metres in length and weigh in at over 90 tonnes – right whales are docile creatures. If their movements were to be included on an Indigenous calendar, it would show that they arrive off the southern Australian coast at the onset of the southern hemisphere winter, having made the long journey from the summer feeding grounds in the Antarctic. Once they arrive, they tend to congregate near the coast, cruising slowly on the surface as they feed.

This behaviour, while wonderful for whale watching, unfortunately makes them easy targets for whalers. So much so that by the mid-1700s, the North Atlantic right whale was essentially extinct for commercial purposes. High in blubber content, and thus extremely profitable, they were the favoured catch of northern hemisphere whalers, who moved progressively south during the 19th century in search of new hunting grounds. The slaughter was relentless, and with the new ships and harpoons, the whales didn't stand a chance.

So effective were the whaling fleets that by the time of Yonge's expedition, southern right whale populations had already been severely depleted. Estimates suggest that over the roughly one hundred years before 1930, over 26,000 southern right whales were taken. By the early 1930s, fewer than a thousand individuals were thought to remain alive, and it was clear that the species was near extinction globally. In a decision of remarkable prescience, the taking of southern right whales was banned by international agreement in 1935. Although recovery has been exceedingly slow – the whales we saw that day off the coast of Albany were three of only roughly three thousand believed to exist – and the southern right whale remains endangered, their protection is an early example of the just-in-time, right-to-the-brink conservation approach that seems to be the way of our time.

After the agreement of 1935, the whaling industry, undeterred, put its ships and processing stations to work on other whale species. The next target in Australian waters was the humpback whale. If you have ever had the chance to see one of these mighty creatures in the wild, it is an unforgettable experience. Compared to the slow-moving, docile right whales, humpbacks are quick and active. They can often be seen within the GBR lagoon, cavorting on the surface, slapping the water with their great flukes, or propelling themselves up out of the water and then breaching back down in huge towers of spray. And while this makes them great fun to watch – whale watching being a relatively modern pastime – it made them harder to catch.

Adult humpback whales grow to 18 metres in length and can weigh up to 40 tonnes – that's as much as six full-grown African elephants or ten average automobiles. Southern hemisphere humpbacks spend most of their time feeding on krill in the open waters of the Southern Ocean during the austral summer, and then migrate thousands of kilometres north to warmer waters to breed in the southern winter. From June to August, they transit the coast of Australia, the males filling the ocean with their complex, mournful whale song, while mothers and calves rest and feed in quiet bays along the shore. During the time of Yonge's expedition, tens of thousands of humpback whales plied the waters of the Great Barrier Reef lagoon each year, and sightings were common.

But despite their power, speed and majesty, and the huge distances they can travel, they were no match for the modern whaling fleets of the 1930s, 40s and 50s. In 1953, the Cheynes Beach Whaling Company in Albany – the owners of the whaling station I toured in 2010 – alone produced over 500 tonnes of whale oil from fifty humpback whales.[3]

By 1954 production had doubled, producing oil worth an estimated £100,000 (an equivalent of over A$5 million today). It was a lucrative business, and business was booming.[4]

Drawn by the huge profits, whaling intensified during the 1960s. By the time I was born in 1962, humpback catches were reaching record levels. But then, the inevitable happened. Too many whales were being killed too fast, and numbers started to decline rapidly. Humpback whales take at least seven years to reach sexual maturity, and the gestation period is twelve months, at which point a single calf is born. Mothers spend up to a year with their calves, teaching them how to survive and protecting them from predators. In this way they are not too different from human babies, completely dependent on their mothers for survival during those early months. Whales were being killed faster than their slow reproductive rate could allow for replacement. Mothers were taken, leaving defenceless calves to perish on their own, or be harvested for what little blubber they could provide.

Soon, the humpback population was in freefall. During those few decades after Yonge's expedition, over 40,000 humpback whales were taken in Australian and New Zealand waters, almost completely wiping out the species. And once again, at the very last moment, some degree of sanity prevailed. Whaling of Antarctic humpbacks ceased in 1963, and the International Whaling Commission protected the species worldwide in 1965.[5]

The recovery of humpback whales since then has been one of the world's great conservation success stories.[6] Surveys in 2021 reported Australian populations had recovered to about 90 per cent of their pre-whaling levels on the west coast, and over 60 per cent on the east coast. So much so that the Australian Government has deemed that the humpback whale no longer requires protection under the *Environment Protection and Biodiversity Conservation Act* 1999, the nation's peak conservation legislation.[7] If we give nature a chance, take the pressure off, be it for whales or coral reefs – and if we don't act too late – it can recover.

The story of whaling in Australia, however, was not over. With so much capital tied up in ships and processing plants, with global demand for their products still active, and hundreds of jobs tied to the industry, there was too much at stake to simply stop. With humpback whales protected, the industry once again simply moved on. The new species in the harpoon crosshairs of Australian whalers was the sperm whale.

Adult male sperm whales grow to about 20 metres in length, and weigh in at around 40 tonnes – about the size of a full-grown humpback. Sperm whales are the largest of the toothed whales, and have the largest brain of any animal. Before whaling began in the 1700s, it is estimated that there were over 1.3 million sperm whales in the world's oceans. In Australian waters, they can be found along the south-west coast, and their range extends down to the Antarctic.

In the second half of the 20th century, sperm whale oil was particularly prized as a specialty lubricant because of its extreme pressure-resistant properties. It was used in mechanical transmissions, high-speed machinery and precision instruments such as custom watches. It was also used in more traditional ways, in cosmetics, toiletries, and to make soap.

Sustained demand for whale oil through the post-Second World War years and into the 1970s meant that whaling remained profitable. Australian statistics are sketchy, but globally it is estimated that by then, sperm whale populations had declined to less than 10 per cent of their original numbers. The Cheynes Beach station near Albany caught and processed a total of 14,695 sperm whales during its operational life, accounting for most of Australia's take.[8] In 1978, it was the last station in Australia to close, ending 178 years of whaling in the waters around

Albany. In 1979, Australia formally adopted an anti-whaling policy, and since has been active in protecting whales globally.

~

The problem with the conservation approach used in Australian whaling, and now in coral reef preservation, is that we can never be sure where the line between 'just in time' and 'too late' lies. It was a debate that was raging in 2019 within our team as we raced to complete the feasibility report into reef restoration and adaptation, due at the end of the year.

When I was growing up in Canada in the 1960s and 70s, one of the things we all knew was that we lived in a country blessed in natural resources. We were, quite literally, the lucky country of the north. I remember sitting in my high-school social studies class, listening to our teacher talk to us about the Atlantic cod fishery. This, he told us, was the single richest fishing ground in the world. Early European settlers to Newfoundland, Canada's easternmost province, spoke of bounty beyond imagining, of so many fish so close to shore that you could scarcely row a boat through them. Throughout the 18th and 19th centuries, cod became the foundation of Newfoundland's economy. New techniques were developed to catch more fish faster. Larger, more expensive boats allowed those who could afford them to travel further in search of cod. By the mid-1800s, near-shore fisheries were becoming exhausted, and the fleets moved north along the coast. By the time of Yonge's expedition to the GBR, the Labrador coastal fishery was gone.[9] New stocks were needed to keep the economy going. The Newfoundland Government urged fishermen to move offshore to harvest the untapped riches of the Grand Banks. Subsidies were provided to help buy larger, steam-powered boats, and ever bigger, more efficient gear.

By the time I was born, in the 1960s, the technological advancements brought to bear on the Grand Banks included the use of sonar to locate cod schools, and huge factory freezer ships that could stay on station, pulling out fish for days at a time, leaving only when their holds were bursting. Ships from dozens of nations trawled for fish just outside Canada's territorial limit. But the sheer scale of the effort now meant that the trawlers were pulling up other fish along with cod, fish that could not be sold. These included huge numbers of capelin – one of the cod's key prey. This so-called bycatch was separated out and dumped overboard, dead and wasted. The cod catch peaked in 1968 at 810,000 tons, about three times more than the maximum yearly catch achieved before the arrival of the super-trawlers.[10]

By the time I was in high school, this wealth supported an entire industry employing tens of thousands of people and generating millions in tax revenues. Our teacher showed us part of a TV documentary about it. We watched as the massive nets were drawn up from the sea, filled to bursting with thousands of quivering silver cod, separated and sized and then dumped live into the holds of factory freezer ships. I remember having nightmares about it for days after, haunted by the images of so many living creatures packed together in hopeless desperation.

That was in 1977. I was fifteen. By then, evidence that the cod fishery was in decline was undeniable. The Canadian Government had expanded its territorial limit from 3 miles to 200 kilometres, taking more control over the fishery, and excluding foreign fishing fleets. But any positive effect this might have had on fish stocks was lost by the poor management of the newly expanded economic exclusion zone. Canadian fishery managers overestimated the health of fish stocks and provided quotas to Canadian and American fishers that were unsustainable. Locals soon noticed alarming declines in fish numbers and size and tried to warn the government. So many fish were now being caught that the surviving stock was unable to replenish itself. Even though few at the time believed it, the population was in terminal decline.

But, as with so many similar stories, there were simply too many reasons to keep fishing, and to believe that everything would be alright. Thousands of people depended on fishing for their livelihoods. Governments relied on the industry for tax revenues. Companies large and small had been built to support the fishery. Deniers pilloried those who called for drastic cuts in quotas as doomsday merchants, economic vandals, and worse. They attacked the credibility of the scientists calling for caution. It worked. Harvesting continued, peaking in the late 1970s, but then declined rapidly.

And then in 1992, the fishery collapsed entirely.

The Canadian Government finally announced a moratorium on cod fishing, but it was too late.[11] Spawning biomass of cod had decreased by at least three-quarters from historically sustainable levels in all stocks, and by 99 per cent in the case of northern cod, previously the largest cod fishery in the world.[12] A fundamental lack of understanding of the ecology of the Atlantic cod was one of the key reasons for the mismanagement. The vital importance of the link between research and management was laid bare.

The immediate effect on the people of Newfoundland was devastating. Over 35,000 fishers and fish plant workers lost their jobs. Hundreds of coastal communities lost their major, and in many cases only, source of employment. So severe was the economic impact that the Canadian Government was forced to put in place a special support package to retrain workers and help the industry switch its catch to other species.

Since the 1992 moratorium, the cod have been recovering, but much more slowly than hoped. By 2010, cod stocks had only reached 10 per cent of the original level.[13] And now a new threat has emerged. Warming of coastal waters due to climate change threatens to further disrupt the delicate balance of conditions that created the bounty in the first place.[14]

The story of the collapse of the world's richest fishery is a tale of warnings ignored, scientific evidence dismissed and opportunities wasted. How could such an amazing renewable resource be allowed to disappear? It is the greatest irony that the reason not to dramatically reduce catches was primarily an economic one. No one wanted to make the tough calls because of the effect on jobs and revenues. Each justification, each manipulation of the facts, each warning from the scientists ignored, built an increasingly flimsy scaffold around the edifice. And, of course, in the end, it all came crashing down. All the jobs were gone, all the revenues and profits disappeared.

If managed sustainably, the fishery could have provided livelihoods and a secure source of high-quality protein indefinitely: just what the world needs, as it struggles to feed more than eight billion people, with more coming every day. The gross mismanagement of the cod fishery in eastern Canada is a poignant example of what happens when politics and greed overrule science, and when the good of the many is usurped in favour of profits for the few. It could happen anywhere, and unfortunately, as I was learning, it was.[15]

~

I didn't have to look far for another example of the profound effect of the 'there for the taking' mentality on Australia's oceans. Chances are that at some point on their journey to Australian in the 1920s, Maurice and Mattie Yonge would have had a taste of southern bluefin tuna. By that time, this magnificent fish was already a delicacy worldwide, and a small line-caught industry was operating in eastern Australian waters. Huge schools of bluefin, numbering in the tens of thousands, were routinely spotted along the continent's southern coast. Again, by all accounts, the abundance was staggering.

These massive, muscular fish can grow to over two metres in length, weighing in at over 200 kilograms. In many ways they are an Australian species, spending much of their lives in Australian waters. By the 1950s, full-scale commercial fishing for southern bluefin was well underway in New South Wales and South Australia. Again, new technology was being brought to bear, in this case the introduction of the purse seine method. A large net is laid out in the shape of a circle around a school of tuna, and then gradually closed up, trapping the fish inside a giant purse. As the ships and nets got bigger, larger schools of tuna were caught. By the early 1960s, mirroring what was going on in my native Canada, the global catch was already peaking at something around 80,000 tonnes a year. The Australian catch didn't peak until 1981, when it too started a long, inevitable decline.[16]

By the 1990s, commercial fishing worldwide had become a loss-making enterprise, in every sense. Not only had specific fisheries been fished to near extinction, but increasingly it was being done at a loss. In 1998, UN food and agriculture economists estimated global fishing revenues at about US$70 billion, and the cost of pulling out that fish at $124 billion.[17] Huge government subsidies for fishing around the world meant that we were actually subsidising the destruction of our fisheries, to the tune of billions of dollars a year. This, surely, is the real definition of insanity.

By 2009 the Australian and global catches of southern bluefin tuna had declined to less a quarter of their respective peaks, and alarm bells had started to peal.[18] Thankfully, for southern bluefin tuna at least, the warnings of the 1990s and the early 2000s were heeded. In 2009, the Commission for Conservation of Southern Bluefin Tuna placed limits on the catches of its member nations, including Australia.[19] Slowly, populations began to stabilise and then recover. Current stock assessments in Australia suggest that the total reproductive output (TRO) for the species is about 20 per cent of the unfished level. Authorities have set a goal of achieving a TRO of 30 per cent of the unfished level by 2035, even though this is still considered to be below the threshold where maximum sustainable yield can be achieved.[20] Thanks to this small recovery, southern bluefin have now been upgraded from critically endangered to endangered by the International Union for the Conservation of Nature. This classification, however, still means that the species remains at a very high risk of extinction.

Relatively low numbers of spawning-age fish, and a dearth of bigger, older adults, means that the population is particularly vulnerable to variations in conditions or unexpected shocks. Climate change is now providing just such a change, on a massive scale. How this will affect the once mighty bluefin we simply do not know. It is this uncertainty that has scientists worried.

The echoes of the warnings of Banfield and Hedley a century ago are deafening. The overfishing of Australian whales, Canadian cod and Australian bluefin tuna are examples of valuable natural species pushed to the brink of collapse by unregulated or poorly controlled exploitation. As demand grew, technology allowed us to take more, and do it faster for longer. And without detailed scientific knowledge of how these animals behaved and reproduced, and little knowledge of the health of their populations, we were essentially driving blind. Even as our scientific knowledge caught up, technological improvements and economic imperatives meant that the harvesting continued and accelerated. Eventually the yearly catch outstripped the population's ability to replenish itself, and at that point, it was only a matter of time until the collapse.

With whales, it appears we got lucky, and stopped the killing just in time to allow recovery to occur. With Atlantic cod, it seems that we did not. The fate of the southern bluefin tuna remains in the balance.

~

In Melville's famous novel, the whale wins in the end. But in the decades following Yonge's expedition, this outcome proved to be the exception rather than the rule.

Seeking access to and profiting from natural resources at favourable terms – the behaviour that drives the tragedy of the commons – is called rent-seeking. As Joseph Stiglitz, the Nobel laureate economist, explains, rent-seeking is responsible for many of the massive fortunes which have been compiled over the last handful of decades. As he says, 'It's far easier to get rich by gaining access to resources at favourable terms than by producing wealth.'[21] In countries such as Australia and Canada where natural resources are owned by the state (and therefore, in democracies, by the people of the state), if access to these resources is provided for free, or at a cost that does not represent their true value, then citizens receive very little benefit from the sale of those resources. Of course, the people benefit indirectly through access to jobs created in those industries, but generally, this income represents a relatively small portion of the overall prize. This type of rent-seeking behaviour is a key driver of inequality in society, concentrating wealth in fewer and fewer hands.[22]

To prevent the collapse of natural resources, one or more of the key drivers of over-exploitation must change. Demand can be reduced by changing public attitudes or developing alternatives, access can be limited by regulation or protection, or those who harvest the resources can be charged appropriate royalties or licence fees for their take, ideally ones that adjust to market prices, allowing the public to share in the windfall when prices rise, and protecting the producer when prices fall. The mechanisms can only effectively be applied and enforced by governments. If they are mismanaged, intentionally or not, tragic outcomes for natural resources and environments can ensue.

The unregulated and unpriced dumping of carbon emissions into the atmosphere is yet another example of the tragedy of the commons, writ on a planetary scale. In a sense we all own the atmosphere – it provides us with one of the most fundamental human rights – the right to *breathe*. And yet, because in law no one 'owns' the atmosphere and the climate it regulates, anyone can pollute it with relative impunity.

It was becoming clear to me that the journey from there to here, from the almost still pristine oceans and Great Barrier Reef of Sir Maurice Yonge's time, to now, was in part a story of the very human notions of infinite abundance and unregulated exploitation.

It is a cliché to call the oceans the planet's last frontier. The word frontier evokes undiscovered territories full of promise and potential. This certainly applies to large parts of Australia's marine estate. Even now, our understanding of the Great Barrier Reef is limited. Despite decades of intensive discovery, there is still so much we have yet to learn. As Charlie Veron put it to me: 'Even now, we know bugger-all about most marine species.'

But the words *last frontier* also suggest places untouched by human hands. In this respect, nothing could be further from the truth. And herein lies one of the most disturbing realities about the current state of the Great Barrier Reef and our oceans more generally. After a century of

A 1933 tourism poster for the Great Barrier Reef. The reef has been a major destination in Queensland and Australia for more than a century, now attracting over two million visitors a year. *Queensland Government Tourist Bureau, 1933, courtesy Queensland State Archives*

The National Sea Simulator facility at AIMS headquarters in Cape Cleveland, Townsville. It is the largest and most sophisticated research aquarium complex in the world. In 2023 construction began on a new phase of the SeaSim, as it is known, which will double its size and capability. © *Australian Institute of Marine Science and Joe Gioffre*

Inside the main experiment room at the SeaSim. The main display tank can be seen on the far left of the image. Many of the long-term experiments on developing heat-tolerant hybrid corals are housed in these tanks. © *Australian Institute of Marine Science and Roslyn Budd*

The author free-diving on John Brewer Reef in late October 2023. This reef is the site of one of the installations of the Museum of Underwater Art, a project designed to promote reef conservation through collaboration between artists and scientists. © *Paul Hardisty*

John Brewer Reef, late October 2023. Corals were healthy, fish and sharks plentiful. By late March of 2024, many parts of the Great Barrier Reef were once again bleaching due to sustained elevated sea temperatures. © *Paul Hardisty*

Maurice (right) and Mattie Yonge on Low Island, 1928. Maurice Yonge, an English marine zoologist, led one of the first scientific expeditions on the Great Barrier Reef.
Photographer unknown, courtesy Russell Album 3, Sir Charles Maurice Yonge Collection, James Cook University Library Special Collections

Low Isle, 1928, taken by a member of the Yonge expedition. The lighthouse can be clearly seen, along with the small barque that brought them supplies from the mainland.
Photographer unknown, courtesy Russell Album 1, Sir Charles Maurice Yonge Collection, James Cook University Library Special Collections

On 18 March 1967 the tanker Torrey Canyon ran aground off the coast of Cornwall in the UK, spilling over 100,000 tonnes of crude oil into the sea. Many in Australia saw the disaster as a warning of what might happen on the reef.

The experimental laboratory on Low Isle, set up and equipped by the members of the Yonge expedition, 1928. *Photographer unknown, courtesy Russell Album 2, Sir Charles Maurice Yonge Collection, James Cook University Library Special Collections*

Exposed coral at low tide, photographed by a member of the Yonge expedition, 1928. *Photographer unknown, courtesy Russell Album 3, Sir Charles Maurice Yonge Collection, James Cook University Library Special Collections*

The Great Barrier Reef from the air is a magical sight. Healthy reefs show dark brown and olive through clear blue water. Aerial bleaching surveys such as those conducted in March 2024 revealed reefs bleached white or fluorescing brightly. © *Australian Institute of Marine Science and Joe Gioffre*

© AIMS | Joe Gioffre

A healthy reef with high coral cover (top), compared with the same reef after bleaching and subsequent coral death (bottom). © *Australian Institute of Marine Science*

research and discovery, we still know so little about them. And yet we are busy changing them as we speak, in ways we have yet to fully observe or understand. As the decades have passed, almost unseen beneath that beautiful, beguiling blue surface, plastics and toxic chemicals have been accumulating and mixing, species disappearing, and whole seascapes changing. And perhaps most worryingly, the oceans are warming, taking up more than 90 per cent of the excess heat produced by the steadily increasing accumulation of anthropogenic greenhouse gases in the atmosphere.[23]

In the 1930s the tourists all wanted to take home a piece of coral for their mantelpieces. Now, people come from all over the world just to see the reef, to feel and experience its stunning beauty, soak up its strangely regenerative aura – the thing that preserved Banfield's mental health and kept him on his little island for the rest of his life. It's so big, and so much of it still visibly untouched, that one can still believe, for a moment at least, in the *idea* of a last frontier. All of this conflict, the bickering between governments and industries and people of all kinds, is a lot more than a protracted battle over who can take the reef's fish or coral or oil or sponges or sea cucumber, and how much they can take. Yes, the reef is huge, but not so huge that it can exist apart, somehow cut off from the rest of the world, immune to the tsunami of change we have wrought on this planet over the last hundred years. Everything we do, literally everything, affects it. Every choice we make, individually and collectively, will contribute to determining the fate of this place. It's not just about how much fertiliser farmers along the coast use on their fields, or how they control their runoff. It's not just about sediment runoff from mining operations, or harbour dredging along the Queensland coast, or how much coal we burn. I was coming to believe that the fate of the reef, and of reefs around the world, and the broader ocean ecosystems they support, and by extension the health of this whole singular place we call Earth, are all tied together, linked in the most fundamental way.

10

Damage Control

In response to the mass bleaching events of 2016 and 2017, the Queensland Government announced that it would introduce legislation in 2019 to improve water quality on the GBR. Progress towards targets was too slow, they said, and the current voluntary approaches were insufficient to make the kinds of improvements the reef now so desperately needed. The reef regulations package, as it would come to be known, was designed to complement existing efforts to improve water quality on the reef. The measures were to include more stringent nutrient and sediment targets for the reef's catchment waters; agricultural standards that targeted nutrient and sediment pollution; and new ways for producers to meet regulatory requirements through accreditation by best management practice programs. Controversially, the package was also to include requirements for farmers to keep records demonstrating their compliance with the standards. The draft bill was based in part on the science contained in the 2017 Consensus Statement on reef water quality – the document I had ploughed through in my first months in the AIMS role.

The reaction to the proposal was instantaneous. Farm groups up and down the Queensland coast were enraged. The new regulations would put additional burden on a sector already struggling with rising costs. Proposed minimum standards would eat into profits, inhibit growth of farming, and might force some farmers out of business. Farming groups like the Australian Cane Farmers Association supported continuous improvement of farm practices, and wanted to do their part to protect the reef, but they favoured the voluntary approach already in place over a regulated system.[1] Canegrowers, the peak body for Australian sugar cane growers, representing around 80 per cent of growers, was even more categorical. While they supported efforts to 'better manage the interactions between farming, water quality in the catchment, and the health of ecosystems within the reef lagoon', and had 'consistently demonstrated a strong commitment to improving practices for both productivity and sustainability benefits', they dismissed the newly proposed regulations as 'a clear example of how not to ensure evidence-based regulation'.[2] They, too, questioned the utility of regulatory approaches, suggesting that the voluntary programs needed more time to deliver results, and more support from government. And tellingly, they criticised the 'general reluctance to embrace and communicate uncertainty in the science of water quality and the reef', claiming that this was due to pressure on researchers to reach consensus on research findings, driven by policymakers keen for clear-cut results justifying their predetermined positions. They claimed that many growers had lost confidence in reef science, and questioned the integrity of the scientific process.[3] A campaign to force the Queensland Government to withdraw the new regulations was launched, with the support of opposition politicians.

Once again, the approach to countering unpalatable regulations or restrictions was to attack the underlying science. Unsurprisingly, Peter Ridd, now a favourite of right-leaning commentators and media, was brought in as the scientific expert to back up the claims that the science behind the new reef regulations couldn't be trusted. He embarked on a roadshow, paid for and promoted by the opponents of the legislation, giving town-hall presentations in farming communities all along the Queensland coast, from Ayr to Cairns.

The lectures were streamed on Facebook, and remain available on YouTube.[4] Despite myself, I watched one.

Proceedings were opened by the chair of Canegrowers Bundaberg, who welcomed everyone and thanked the Australian Environment Foundation (a spin-off of the Institute for Public Affairs) for financing the lecture tour. He then handed over to Tom Bostock, chair of the AEF, to introduce Ridd, 'an acknowledged expert on the Great Barrier Reef'. Bostock began shakily, but soon settled in, reciting the organisation's mission almost word-for-word from its website. 'The AEF,' of which Peter Ridd was a director, 'is a non-profit, membership-based organisation. It seeks to protect the environment, while preserving the rule of law, property rights, and the freedom of the individual that underpin the material progress that is required to do so.'

'We take an evidence-based, solution-focused approach to environmental issues,' he continued. 'While it may be true that we are all environmentalists now, the great majority of Australians have little or no say in the environmental policies that are most often proposed to Australian governments – federal, state or local. These policies are almost exclusively the domain of a tight network of ideologically-driven environmental groups ensuring one view, and one view only, is put forward.'

In happy contrast, the AEF was 'non-ideological, solution-based, and puts a premium on good science and rigorous evidence'.

After a brief plug to the audience for new members, Bostock went on to describe the shabby way Ridd and his mentor Bob Carter, now deceased, had been treated by James Cook University, and Ridd's subsequent successful court action against JCU. Ridd sat next to the podium, eyes downcast, as if in quiet contemplation of all he had suffered.

Then Ridd stood at the podium and looked out at the audience, dozens of country folk gathered into a town hall, squinting behind his glasses. He was dressed for the occasion in a crisply ironed blue-collared shirt. His greying beard was neatly trimmed. He looked every inch the seasoned, credible, reasonable scientist. The first slide was put up, and he began to speak. The title of today's talk was 'How Reliable is the Science Demonstrating Damage to the Great Barrier Reef?' In a high-pitched voice, he launched into his now-familiar story.

The content was simplistic, the slides crudely made. There was a major crisis in science overall, he said. The peer-review system to ensure the quality of scientific publications was a sham, often involving nothing more than a quick read of a few hours by a couple of colleagues. It wasn't long before he was on to his favourite subject, intended to show how peer review doesn't work: a relatively arcane paper in the prestigious peer-reviewed journal *Science* about coral growth rates, written by Glenn De'ath in 2009. In that study, De'ath and his colleagues found a 15 per cent decline in coral calcification rates in massive *Porites* brain corals between 1990 and 2005.[5] Back in 2013, Ridd had found two errors in the paper, which were subsequently corrected by the authors and published that same year, with a revised decline of 11 per cent.[6] Ridd didn't mention that correction, or show the revised data. He stuck to the original, and then showed his own graphs, which of course showed no decline in growth rates. It was a theme he would return to again and again over the coming years, himself and through others: peer-reviewed science was untrustworthy.

But in using this example of coral growth rates, he again seemed to ignore that peer review was about exactly this – the open publication of scientific work that is then open to scrutiny, re-analysis, and challenge by thousands of scientists across the world. In this process there is no room for triumphalism. It isn't about 'gotcha' moments, nor about expecting that a single paper written more than a decade before, subsequently revised and updated, should continue to be examined as if it existed in a vacuum. Science is about building on existing knowledge, step by step, testing hypotheses, learning from errors and oversights, and moving ahead.

He kept going. By now I had overcome my initial frustration and was losing interest. He talked about sediment movement to the reef, a subject he actually knows a lot about (he did a lot of his good work, before moving into advocacy for the AEF and the IPA, on this subject), again claiming that farmers were not affecting the reef. He disputed claims that the reef was badly bleached in 2016, claiming only 8 per cent of the reef was affected, but provided little if any scientific basis for this conclusion. Coral cover had actually jumped 250 per cent in the southern reef in less than ten years, he claimed. But it was hard to follow because he provided no time frames – 250 per cent from when to when? All of this in the name of exposing what had now been upgraded, midway through the talk, to '*extremely* untrustworthy science'.

Then the attacks started. A photo of Ove Hoegh-Guldberg, the eminent University of Queensland coral reef scientist, came up on the screen. Hoegh-Guldberg was ridiculed for pointing out that peer-reviewed science is also used to develop the technology that keeps our planes in the air. The next slide showed an old biplane nose down in the dirt. 'A plane made by a GBR scientist,' chortled Ridd.

He was hitting his stride now. The words flowed easily. Coral cover on the reef hadn't changed. Growth rates had not dropped, and if anything, were increasing. The reef was 'absolutely fabulous'. AIMS and other reef scientists (except him) were producing untrustworthy results. Half of it was likely to be wrong, but he didn't know which half. 'A lot' of the 2017 Consensus Statement on water quality, the product of years of effort by dozens of Australia's best scientists, was just plain wrong. Which part he didn't say. Presumably he knew but he wasn't telling. It didn't matter. The audience loved it. He was telling them what they wanted to hear: they, the farmers, were not affecting the reef. They were being unfairly blamed and targeted. The new reef regulations should be scrapped. He closed to rapturous applause.

The lecture series to farmers soon attracted the attention of the Independent Expert Panel on the Great Barrier Reef, chaired by ex–chief scientist Professor Ian Chubb. In a letter to federal environment minister Sussan Ley and Queensland premier Annastacia Palaszczuk, the IEP stated that it could not sit by and watch as reef science was misrepresented, likening Ridd's approach to the tobacco industry's attempts to spread doubt about the effects of smoking.[7]

It wasn't long before I was fielding direct enquiries from farmers. I answered every letter and phone call personally. Most were straightforward and polite. One cane farmer, Paolo (not his real name), invited me to his farm, south of Townsville. I drove out on a warm sunny day and met him on a side road, just off the highway. We shook hands, looked each other in the eye. He was deeply tanned, dark-haired and built like a rugby league player, with broad shoulders and powerful forearms.

Over the next few hours, Paolo led me on a tour of some of the farms in the area, showed me the various improvements that had been made over the years to capture runoff and prevent harmful chemicals from getting to the water courses and out to the reef. We met one of his neighbours and were shown his computerised monitoring system which provided detailed up-to-the-minute information on water flows and fertiliser use. Farmers have to pay for fertiliser, he explained, so it was in their best interest not to use any more than they absolutely needed to maximise yields. I was impressed, and I told him so.

Eventually, the question of the new regulations came up. He asked, why were we blaming the farmers? Farmers care about the reef, he told me. This was their home. They fished and holidayed on the reef. They were invested. I asked him why they were blaming the scientists. I tried to explain that we were not the ones who made policy. How the government used our science was up to them. It sounded like a cop-out, but it was true.

Before I left, he took me to watch one of the workers setting the cane alight – the burning that takes away the dried-out husk, leaving only the valuable cane. Paolo explained that this made harvesting easier and more efficient. He asked if I would like to have a go. I nodded. He checked the wind direction and showed me how to use the flaming oil can. He pointed to a place. Start there, work to your left. I put on the gauntlets, strapped on the can and moved to the spot.

As soon as I set the wand to the cane, flames burst up from the stalk. I jumped back in surprise, recovered and started moving left, lighting as I went. Soon an inferno two storeys high was roaring through the field. The heat was so intense I was knocked back, as if someone had straight-armed me in the chest. I gasped for breath. Afterwards, we posed for a few photos in front of the flames, shook hands and agreed to stay in touch.

As Ridd spread his message up and down the coast, local politicians opposing Queensland's Labor government came on board. A group of One Nation and Coalition senators moved in the upper house for a formal inquiry into the question of whether farmers were impacting the reef, and if the proposed regulations would hurt farm productivity. On

17 September a public inquiry into the matter was launched, chaired by WA Labor senator Glenn Sterle and co-chaired by Queensland LNP senator Susan McDonald. Other members of the committee included Senator Malcolm Roberts from Pauline Hanson's One Nation party, and the LNP's Gerard Rennick, both from Queensland. Submissions were invited from federal and state government departments, industry stakeholder groups, community groups and individuals.

In total, the committee received 120 submissions from a wide variety of groups and individuals. The AIMS submission included a detailed summary of the science linking runoff from the land with deteriorating water quality on the reef, especially in the inshore parts of the World Heritage Area.[8] Public hearings were scheduled for March 2020.

Undeterred by the growing clamour, the Queensland Government passed into law the arcanely titled *Environmental Protection (Great Barrier Reef Protection Measures) and Other Legislation Amendment Act 2019* on 19 September. The measures were to take effect on 1 December 2019.

That year, officially Australia's hottest and driest to date, closed with Australia on fire. The PM, who had slipped away for a family holiday in Hawaii, was forced to return as the country suffered through the worst bushfires in its history. Over ten million hectares across Southern Australia had gone up in flames, more than the 2009 Black Saturday and 1983 Ash Wednesday fires combined. Thirty-three people died, and some estimates suggested that over a billion animals perished. Scientists from the CSIRO were unequivocal. Without climate change, fires of this magnitude, intensity and extent would not have been possible.[9]

In a paper in the highly respected journal *Science*, Ove Hoegh-Guldberg and colleagues from around the world concluded that it was now 'imperative' for the human race to stabilise warming at 1.5 degrees

Celsius above pre-industrial levels.[10] The authors concluded that the risks to humans and the planet's ecosystems of continued warming were severe, and worsening with every tonne of carbon released into the atmosphere. Reprising Stern's message of more than a decade before, the benefits of action would be significantly greater than the costs of acting. Doing nothing was not an option.

As if to underline the point, the GBRMPA released its 2019 Outlook Report. The language in the report was direct and unequivocal. 'Climate change is the greatest threat to the Great Barrier Reef,' it stated. 'Only the strongest and fastest possible actions to decrease global greenhouse gas emissions will reduce the risks and limit the impacts of climate change on the Reef.' The report cited the growing recognition that limiting global average temperature increases to 1.5 degrees Celsius or less was 'critical to minimise significant environmental and societal costs from the loss of reef habitats'. The outlook for the reef was downgraded to 'very poor', signifying that the integrity of this World Heritage property was increasingly in question, and that its sheer size was no longer a guarantee of resilience. The biggest threats to the reef were climate change and land-based runoff. 'Given the current state of the Region's values, actions to reduce the highest risks have never been more time-critical.'[11]

Meanwhile, we were hearing the first reports from Italy of a strange and deadly sickness that was spreading quickly, killing off the elderly and those with underlying health issues in alarming numbers. In the end, the emergence of the Covid pandemic would cause the Senate hearings to be postponed until July of the following year. And the opponents of the new Queensland farming regulations would put the extra time to good use.

~

In Margaret River, on the south-west coast of Western Australia, the second day of the new year of 2020 dawned sunny and warm. I was looking forward to a day of kayaking and walking a section of the beautiful Cape to Cape track. Our gear was packed and we were about to start walking down the hill to the river when I made the mistake of checking my phone.

There was an urgent message from John Liston, our communications manager at AIMS. 'Call me immediately.' I could see Heidi quietly fuming as I lifted the phone to my ear. I knew that look – remembered a sailing holiday back when the kids were small. I'd spent a large part of the first four days of that trip on the phone dealing with a crisis at work. On the fifth morning, the phone was gone. Disappeared. I could have sworn I'd left it on the chart table, as I did each night, ready to be used in an emergency. Over the next day I tore that boat apart, but never found my phone. Of course, when I got back to work two weeks later everything was fine. The world, miraculously, had kept turning without me. Officially, the mystery of the lost phone was never solved.

'Happy new year, John. What's up?'

'Yeah, happy new year. Check your news feed. Our friends are at it again.'

'Will we need to respond?'

'Yes. I think we will. See what you think.'

'I'll call you back.'

At this point, I was expecting the worst. Sea surface temperatures on the GBR had been unusually high over the last two months of summer, building up a lot of accumulated heat. This despite La Niña conditions in the Pacific that would normally be expected to bring cooler conditions. Signs of bleaching had already been observed on some reefs off Townsville, and we were bracing ourselves for more bad news.

I opened my email, scrolled down, found the alert and opened it. An article in *The Australian* – the same newspaper that, back in the 1970s, played such an instrumental role in saving the Great Barrier Reef from oil drilling. This one was by their environment editor Graham Lloyd, entitled 'Reef coral testing "flawed, needs fix".[12] The short article seemed to cover much of the same ground as Peter Ridd's coral growth-rate story, and reached similar conclusions about reef science that had featured in Ridd's lectures to farmers and his opinion piece that featured in the same 2 January 2020 issue of *The Australian*.[13] AIMS was again mentioned by name, as was the now retired Glenn De'ath and his 2009 paper, but again not the correction paper of 2013.

I called John back right away.

'Hi John, I've read Lloyd's piece.'

'What do you think?'

'He's repeating and amplifying Ridd's message.'

'They're giving him a platform.'

'And a megaphone.'

'So, what do we do?'

We had already responded in the scientific literature, pointing out clearly the inaccuracies of these claims. That was the right place to have these discussions – the peer review process working as it should, providing a place for reasoned and fact-based discussion and evaluation of the value and validity of scientific work. But this wasn't just a Facebook or YouTube video – this was one of the largest daily newspapers in the country, with hundreds of thousands of readers. And that unsubtle message about the untrustworthiness of reef science, reef scientists, and reef research institutions was, via *The Australian*, now getting much wider coverage. We had to respond. We decided that Dr Britta Schaffelke, head of our Great Barrier Reef research program, should correct the record.

Later that day, walking along the bluffs south of the river mouth and looking out over the azure waters of the Indian Ocean, Heidi and I spotted a pod of dolphins playing in the surf. We stood watching them for a long time.

The next morning, I got another call from John. He had contacted Graham Lloyd and discussed our concerns with him, and sent him our response to the article, correcting the record. *The Australian* published an editorial the same day on the $443 million grant to the GBRF, taking up Peter Ridd's call for some of the grant to be used to check reef science.[14] But this new year's offensive wasn't over yet. On 4 January, Graham Lloyd followed up, this time with the headline 'Great Barrier Reef Foundation "spending millions on costs".[15] The article revealed that 'one-third of expenses for the Great Barrier Reef Foundation in the most recent reporting period was spent on administration and fundraising. In total, almost $50m will be spent on administration between now and 2024.' The article included a statement from the Great Barrier Reef Foundation saying that it was on track with Australia's largest environmental funding campaign, and that the 'target is to grow the federal government's $443m contribution into $800m to invest into research and projects'. But the story quickly moved to a reprise of the AIMS coral core issue, again placing both issues side by side in the minds of readers. Lloyd wrote, 'The cost of administration for the Reef Trust, came as the Australian Institute of Marine Science has defended research showing an unprecedented collapse in coral growth rates but said the findings did not attempt to predict the future.' This time, to Lloyd's credit, he included Britta's response: findings of a decline of coral calcification rates from 1990 to 2005 were valid and supported by other studies in similar reef regions around the world. The detailed peer-reviewed re-analysis of 2013 confirmed a decline, this time of 11.4 per cent rather than the original 14.2 per cent. Still a decline. And further studies in 2014 demonstrated clearly that coral bleaching on the GBR in 1998 and 2002 had supressed growth rates for four years.

Still, I couldn't help wondering why Lloyd had chosen to end a story about the controversial GBRF funding with a reprise of Ridd's critique of AIMS coral growth-rate work of more than a decade ago. By now the Great Barrier Reef Foundation was battling against a wave of largely unwarranted attacks on its reputation, but had unquestionably been tainted by the government's handling of that massive grant. Whatever the intent, my view was that just being mentioned in the same article, under that headline, might confuse some readers. We had already received several online comments confusing us and the GBRF, stating that AIMS had 'all this money'. Sure, the statements about the GBRF were true. After receiving so much money, the foundation was in a phase of building up its capacity to administer the grants effectively and transparently. That phase would take some time yet to complete, and in the meantime, administrative costs would be proportionately large. Over the life of the grant, the GBRF had already publicly agreed to cap administration costs at 10 per cent of the total, well within the range of other government programs. And yes, AIMS did defend, through the peer review discussion process, the coral growth work republished in 2013, and yes, coral cores are a window into the past and not a crystal ball to the future. But more importantly, science moves on. Work done more than a decade earlier, then refined a few years after that, is just one more step in building up our still imperfect knowledge of one of the most complex ecosystems on the planet. But whether fairly or not, the GBRF was being hammered by commentators from the left and the right. Associating AIMS with that controversy was, in my opinion, another subtle way of continuing the campaign to undermine reef science.

It was now clear that a coordinated effort was underway to discredit reef science and associate it with deeply unpopular government decisions. The Morrison government, still less than a year in power, was deeply antithetical to meaningful action on climate change, and with the reef now showing signs of recovery, conservative commentators were going on the offensive.

~

By early February of 2020 it was apparent that something worrying was unfolding around the world. The illness that was now being referred to as Covid-19 had spread from China to Europe and was starting to take hold in the United States and Canada. Highly infectious and unpredictable, the SARS-CoV-2 virus attacked the respiratory system, quickly overwhelming the immune system. Here in Australia, we watched nervously as the first cases were reported in Victoria and Sydney. The first international travel bans were put in place, preventing movement between certain countries, but it was too late.

By March, infection rates around the world had skyrocketed, and people were dying. Lots of them. Every night the TV news showed pictures of overwhelmed hospitals and morgues choked with bodies. On 11 March, the World Health Organization declared a global pandemic. On 20 March, with cases rising in Australia, the government's new National Cabinet decided to close the borders. Anyone coming into the country would have to quarantine for two weeks. Several states followed suit, preventing movement across their borders. Australia, unlike many other countries, decide to pursue a zero-Covid policy until vaccines became available and enough people were inoculated to protect the population.

Every night on our TVs, the battle between science and superstition, between truth and conspiracy, played out before us. It was all there, in real time, from the plainly outrageous to the subtly idiotic: Donald Trump suggesting that ingesting bleach would cure Covid; claims that the virus was being spread by 5G towers; assertions that masks didn't work; brave predictions that it was nothing more than a flu and would be over soon. We watched Dr Anthony Fauci duelling with President Trump, held our breath as economic indicators crashed and millions were thrown out of work.

As the head of AIMS, I had some big decisions to make. Our people were worried about their health, and nervous about their jobs. At that stage there was no way to tell who was infected and who wasn't. Some people, apparently, could have Covid, be infectious to others, and yet show no symptoms. Our work depended to a large extent on being able to put people on ships and under the water, and on the 24-hour-a-day operation of our laboratories and the Sea Simulator. We'd seen what had happened to the cruise ship industry, how vulnerable a group of people were when crammed together for long periods of time on a boat. If one person was sick, everyone would get it. Our emergency management committee was working overtime to develop a plan, coordinate with other government agencies, get those of our people stuck overseas safely back into the country, and communicate what was happening to staff.

In mid-March, I called our staff together using Microsoft Teams – like everyone else we were transitioning quickly into online work – and told them that whatever happened, no one would lose their jobs. One way or another, we would keep everyone on. Despite what was going on, there was a lot of work to do. We would follow the scientific advice – we were scientists after all – and do everything we could to protect each other. Our people were urged to look after themselves and their families first. 'If in doubt, stay out,' became our motto. One of our scientists, a microbiologist, gave a presentation on the current state of knowledge of the SARS-CoV-2 virus and its transmission. Medical science had found solutions to so many of humanity's great afflictions in the past, and would do so again. Until then, we would set up rostered isolated work teams for Sea Simulator staff, we would work from home whenever possible, socially distance, avoid crowds on our days off and use masks. And we would keep working.

At the same time, reports were coming back from our teams on the water that coral was bleaching extensively in reefs from Cairns to Townsville. It was now apparent that another mass bleaching event was unfolding on the reef. That this was happening during a La Niña year was profoundly worrying to our scientists and to our colleagues at the GBRMPA. To Terry Hughes of the Centre of Excellence for Coral Reef Studies at James Cook University, this was a clear sign that climate change had shifted the temperature baseline up enough that, depending on local conditions, bleaching could now theoretically occur in *any* given year.[16] This represented a fundamental tipping point in the history of the reef, although few could see it at the time.

Worried like everyone else about Covid, dismayed about what was happening on the reef, Heidi and I kept pretty much to ourselves. We went into town only for food, returning to the relative isolation of our house on the peninsula. Evenings we walked the AIMS beach, reflecting on the calamity befalling the world.

'It's nature fighting back,' said Heidi one evening as we approached the northern end of the beach, the rocky headland already in shadow. 'There are simply too many of us.'

Viruses jumping from animals to people was one of the emerging theories of Covid's genesis. As humans increasingly encroached on wilderness, buffers between the two broke down, allowing diseases to cross over.[17] Another popular theory was that it was human-made, and had escaped from the Institute of Virology in Wuhan in China, the original epicentre of the outbreak.

'If it is nature fighting back,' I said, 'it's working.'

The death toll globally was already approaching a hundred thousand. Little did we know that by the end of the year the official World Health Organization death toll would be over 1.8 million, and that by August 2023 over 6.9 million people would have died from the disease.[18]

We walked on in silence, chastened by the thought of this revenge. As if to point out our complicity, we passed several recently dug-up turtle nests, the shattered eggshells scattered like frost about the lip of each hole. We counted five more before we decided to turn around and head home.

'It's wild pigs and feral foxes that are doing it.' It seemed the list of our transgressions was limitless.

'We still haven't done anything to stop it,' said Heidi after a while.

'No.'

'We should. This year. This nesting season.'

'Okay. No more excuses. We'll set it up. Ask for volunteers.'

She squeezed my hand. 'Good,' she said. 'That's good.'

As the world fell into lockdown, and the global economy crashed, a strange thing started to happen. The world quietened. In India, people who had lived their whole lives without ever seeing the Himalayas because of the permanent pall of air pollution, were treated to awe-inspiring panoramas of ice-capped peaks shimmering under impossibly blue skies. In Europe, wild animals not seen in the region for decades walked deserted city streets. Skies normally woven with vapour trails emerged pristine and eerily silent as whole fleets of airliners were parked. Too often enthralled with our own handiwork, we were forced into silence and given the opportunity to appreciate Creation.

As they were for many people around the world, for me the early months of the pandemic were a time of bewilderment, worry, frantic action and long periods of unfamiliar quiet. Unable to travel, or even visit our favourite parks or wilderness areas, with restaurants, gyms and clubs of all kinds closed, there was suddenly time for reading and introspection. Afternoons, after the emergency management team meetings and the phone calls to other government departments were over, I would leave my office and walk the site.

It was an eerie feeling, wandering empty corridors which only a few weeks ago had been bustling with activity. Vacant offices and blank computer screens, empty chairs and parking lots, the warm afternoon sun streaming through west-facing windows, motes of dust and tiny fibres of carpet suspended in the thick yellow beams. It was as if time had stopped, and with it, human endeavour. We were facing a problem for which, at the moment it seemed, we had few answers. Any hope we had of bringing the pandemic under control and saving lives clearly lay in the application of science, the hard-won methods and tools developed over centuries. Not in hand-waving denial and the empty posturing of ideology.

What we needed now, more than ever, was the rational application of scientific principles and methodology: develop hypotheses based on observation and deduction; test those hypotheses through experimentation and the collection of data; analyse the data to identify trends and patterns, and if possible, causality; and then, develop and test solutions. This takes time, and rigour. What we needed was for Anthony Fauci to win out over Donald Trump. It was if we were fighting the battles of the Enlightenment all over again.

And that was what this place was for. Just another outpost in the long war of rationality over superstition, the human struggle for understanding, for answers. This was a place of science, of knowledge, of enquiry, the buildings and equipment and people, a whole institution created almost half a century before to do exactly that – find answers, shine a light into the dark reaches of the unknown.

11

Layers of Defence

Back in the 1960s, as Wright and Büsst and others fought to protect the reef against mining and oil development, it was as abundantly clear, as it had been to Yonge and Hedley decades before, that to protect the reef, we needed to understand it. As early as 1963, John Gorton, then minister for science, testified at Senate estimates that 'there is justification for the establishment in Australia of a marine biological research station ... By failing to explore the possibilities of the Great Barrier Reef we have not done justice to the scientific world.'

Wright herself wrote that at the time, 'Ecological knowledge of coral reefs was scarce not only in Australia but overseas'. In 1968, Professor Frank, the editor of the respected *Journal of the Ecological Society* in the United States, had written to the Great Barrier Reef Committee, stating that 'we simply do not, at this stage, have major studies on the dynamic interactions between organisms of the highly complex coral community'. He went on to suggest that there was a 'shocking lack of appreciation ... of the importance of the Great Barrier Reef as a major biological phenomenon'.[1] Things hadn't changed that much since Yonge's time.

By 1969, the pressure on governments to address this lack of knowledge about the GBR had become irresistible. One of the main proponents for a new Australian marine research capability was Cyril Burdon-Jones from the University College of Townsville. Burdon-Jones was the university's first chair of marine biology, and a tireless campaigner for increased funding for science. Sensing that the time was right, Burdon-Jones developed a detailed blueprint for what he called a national research centre for tropical marine science in Australia, including a new facility with laboratories, a library, space for twenty scientists and up to fifty support staff, three new boats including a large ocean-going coastal research vessel, and a site on the coast with its own wharf. He went to Canberra and pushed the government to fund his vision. Now was the time.

So, when John Gorton became prime minister in October 1969, Burdon-Jones must have thought that his dreams were about to come true. Gorton would make good on his election promise, and his new centre would be born.

Unfortunately for Burdon-Jones, it didn't work out that way. Politics, as it so often does, intervened. Instead of funding a centre at what by then was known as James Cook University, a Queensland state institution, the Commonwealth Government passed *The Australian Institute of Marine Science Act* 1970 into law, and AIMS was born. Despite being a national agency, the new institute's initial focus was to be squarely on the Great Barrier Reef.

It would take another five years and a change of government for a permanent site for the institute to be chosen, but in 1975 construction began on the buildings I was now working in on Cape Ferguson. Dr Malvern Gilmartin, formerly professor of biological oceanography at Stanford University in the United States, was appointed as the institute's first director that same year. The funding of AIMS crushed Burdon-Jones' grand plans for a marine science centre at James Cook University. By

1975, his department numbered only a handful of staff, but had managed to secure a small field station on Orpheus Island. Still, it must have seemed a cruel irony. The initial scale, scope and mandate of AIMS bore an uncanny resemblance to his own blueprint developed in 1969. Even now, the place remains a monument to his vision.

Finally, after decades of requests, appeals and petitions, the Great Barrier Reef had its own dedicated research organisation and a field station where university researchers and students could study the reef up close. It was a huge step forward, but there was more to come.

The election of Gough Whitlam's Labor party in October 1972 brought in an era of reform in Australia that arguably matched what was occurring in the United States at the time. In 1970, President Richard Nixon, a Republican, created the United States Environmental Protection Agency, with responsibility for conducting environmental research, assessment and education, and with power to create and enforce regulations to protect the country's natural environment, the first body of its kind in the world. This was followed by a wave of new environmental legislation, including the 1970 Clean Air Act, the 1972 Clean Water Act and the 1973 Endangered Species Act. Never before had a country enshrined protection of its natural environment so thoroughly in law. It was breathtaking stuff, and the Australian Labor Party was paying attention.

In 1973, Australia joined the IUCN, which had been established by the United Nations in 1948. Being a member meant committing to following the organisation's guidelines for preserving and protecting wild places and natural resources. About the same time, the Whitlam government set up a committee of inquiry into the national estate under Justice Robert Hope of the NSW Supreme Court. The Hope Report, released in 1974, read like a catalogue of highway accidents.

'The Australian Government has inherited a national estate which has been downgraded, disregarded and neglected,' began its findings and recommendations section. 'All previous priorities ... have been directed by a concept that uncontrolled development, economic growth and "progress", and the encouragement of private against public interest in land use, use of waters, and indeed any part of the National Estate, was paramount.'[2] As with so much that I've read, the words contain an eerie resonance, as if they could have been written today. Mentioning the reef specifically, the report concluded that threats to its continued existence were many, and urged government to move quickly to establish a national marine park to protect it.

Meanwhile, the royal commission launched in 1970 to investigate the issue of oil exploration on the reef was finally ready to deliver its report. After three and a half years hearing the testimony of ninety-five witnesses from around the world, the massive thousand-page report found that little was known anywhere in the world about the long-term effects of petroleum compounds on coral organisms at any stage of their lives. It went on to reprise what was now clear to anyone paying attention: the reef was still a mystery. We simply did not know enough about it, about how the myriad of creatures who made up the 'it' lived and interacted and reproduced and died. In a nod to the new institute being set up on Cape Ferguson by Dr Gilmartin, the report expressed confidence that 'in time', those secrets would eventually be unlocked. Therefore, the commissioners concluded that the risks were real, and while the likelihood of major blowouts was small, the effects could be 'substantial'.[3] This notion of so-called 'Damocles risks' – very unlikely but so catastrophic that if they did eventuate the damage would be unacceptable – is something now well enshrined in modern risk assessment science.

But the royal commission's findings were not universally popular. The hearings themselves were often filled with tension as witnesses were cross-examined, sometimes ruthlessly. The Queensland Government remained glued to its stated policy of 'controlled exploitation' of the reef, extolling the potentially huge economic rewards, and representatives of the Australian Petroleum Explorers Association*, questioned how allowing tourism on the reef was not, in itself, a form of controlled exploitation. Both could cause damage to the reef if not properly controlled and managed. It was simply a matter of putting the right controls in place. Conservationists maintained that an oil spill, however unlikely, would have devastating effects not only on marine organisms, but on tourism, now a burgeoning multimillion-dollar industry.[4] In the end, however, the commissioners found that oil and gas exploration could not be safely conducted on the reef. All drilling should be postponed until the long-term research needed to fully understand the possible impacts was complete.

Despite this reprieve, the reef's future remained precarious, balanced between state and federal jurisdiction, still barely understood, and under increasing pressure from tourism, shipping, and outbreaks of crown-of-thorns starfish. Buoyed by the Hope Report and informed by the findings of the royal commission, the Whitlam government quickly seized on the idea of a specially legislated national park encompassing the entire length of the reef, the vision first articulated by Edmund Banfield all those years ago.

The *Great Barrier Reef Marine Park Act* was passed into Australian law on 22 May 1975, with strong support from both sides of politics. Members queued up to speak in support of the bill. The reef was a

---

* Established in 1959, APEA was later known as the Australian Petroleum Production and Exploration Association, until 2023 when it rebranded as Australian Energy Producers, ditching the word petroleum.

'priceless heirloom', that deserved long-term protection so it could be enjoyed by all Australians. After all this time, it had finally happened. Protected for all time as a national park, the reef, so everyone thought at the time, was safe.

~

Even before the legislation establishing the Great Barrier Reef as a national park was passed in 1975, it was becoming apparent that administering this huge new park would be no easy thing. The royal commission had revealed the depths of feeling among the various groups looking to claim their piece of the GBR's wealth, wonder and mystique. And the legislation had not solved the friction between the Commonwealth and Queensland governments. Several High Court actions were now underway, led by states increasingly hostile to Commonwealth control. Any attempt to manage the park as a single entity would require the cooperation of Queensland. It controlled the reef's islands and the waters within 3 miles of the islands' high-water marks. To make it work, the two governments were going to have to cooperate.

But when Prime Minister Whitlam wrote to Premier Bjelke-Petersen in September of 1974, proposing the creation of a Great Barrier Reef Marine Park Authority, and asking for Queensland's cooperation, the response he received was equivocal. Bjelke-Petersen felt that it was not the time for 'impulsive action', and suggested that 'all other matters should be accorded their just priority'. This wasn't 'unwillingness to cooperate', he stressed, but rather caution on his part, given that the High Court challenge into the *Petroleum and Minerals Authority Act* 1973 had not yet been decided.[5]

Despite these headwinds, the Whitlam government pressed ahead with its plans. The Act was passed and the Authority came into being. Its role would be to determine which parts of the region should be included as part of the new marine park; decide on how different parts of the park were to be used; and from that identify zones for tourist development, fishing, conservation, and shipping.[6] The Queensland Government continued to maintain that the legitimacy of the new park authority was contingent on the outcome of the High Court challenge. Over the next several months it took a go-slow approach, engaging with the Commonwealth and its new authority, but only just.

And then, on 11 November 1975, Remembrance Day in Australia, everything changed. Prime minister Gough Whitlam was sensationally dismissed by governor-general Sir John Kerr. It was the first time in the country's history that the governor-general had exercised his reserve powers and sacked a prime minister who had the support of a majority of members in the House of Representatives. The opposition leader, Malcolm Fraser, was appointed caretaker PM, and a double-dissolution election was called. The Coalition won, and Fraser became Australia's twenty-second prime minister.

With the change in government from Labor to the more conservative Coalition, the situation changed quickly. Within days of the election, Queensland's National–Liberal cabinet agreed to engage with the new park authority, and nominated members to its governing board and consultative committee. And not long after, the High Court announced its ruling on the *Seas and Submerged Lands Act*. The legislation was found to be valid and constitutional, and the Commonwealth's position on jurisdiction over the reef was vindicated. After more than half a decade of constitutional wrangling, the path was now clear, and the Great Barrier Reef Marine Park Authority came into full effect.

Not only was the Great Barrier Reef protected in law as a park, it now could be properly managed and regulated. It was a big win for the reef.

In June of that pandemic year, amost fifty years after those fateful decisions, we kayaked along the eastern coast of Hinchinbrook Island, north of Townsville, camping and hiking the famous Thorsborne Trail. With an area of almost 40,000 hectares, Hinchinbrook is Australia's largest island national park, and has been protected since 1932. The trail was named after Margaret and Arthur Thorsborne, who fought for decades to prevent development on the island. As we walked, I could see clearly the cause and effect of it all. If you open a place up to development, then slowly, eventually, the incursions will take their toll. The resorts, the access roads, the drilling platforms and quarries, and all that other essential human activity must and will leave their mark, and change a place irretrievably. Entropy moves in one direction only, and you can't go back. But if you fight to protect a place, over decades if you must, backed by science and a clear sense of what is right, then wins can be fashioned, small and big. And this was a big one.

~

In the end, it would take Fraser and Bjelke-Petersen until 1979, almost four years after the marine park legislation was signed into law, to reach full agreement on how to cooperate on the Great Barrier Reef. This would lead to the final major breakthrough of this remarkable period for the reef.

In 1972, that same year I was snorkelling for the first time in the Caribbean as a wide-eyed ten-year-old, the world got together in Paris and decided to protect the planet's unique natural and cultural heritage. It was agreed that places like the Grand Canyon and the Pyramids should be preserved from degradation and maintained for future generations. The United Nations Educational, Scientific and Cultural Organization

(UNESCO), proposed an international convention for this purpose, to be known as the World Heritage Convention. Australia signed on in 1974, following the recommendations of the Hope Report.

But progress was slow. State governments in particular were hesitant to have their properties listed because it created a barrier to development.[7] Throughout the country, areas of outstanding beauty and of deep significance to Traditional Owners continued to be developed and degraded. But the recent wins on the reef had come at the right time. Now a national park, there was no reason not to recognise the reef's global significance. The Commonwealth and the State Government agreed: the entire marine park along with all of its islands and Queensland waters would be included in an application for listing as a World Heritage Area.

That the world formally recognised what many Australians had long known to be true should not have been a surprise. Tourism on the reef was growing fast, and more and more of the visitors were coming from overseas. My first visit was in 1975. I was thirteen years old when my father, a lifelong traveller, announced that we were off to Australia for Christmas to visit his best friend, my godfather, in Melbourne. It was the trip of a lifetime. I saw my first cricket match, watched Viv Richards score a century against Dennis Lillee and team at the Melbourne Cricket Ground. We walked the Sydney foreshore and toured the Opera House. But the highlight was snorkelling on the mythical Great Barrier Reef.

To qualify for natural World Heritage status, nominated sites needed to be recognised as meeting at least one of four criteria: contain superlative natural phenomena or areas of exceptional natural beauty and aesthetic importance; be outstanding examples representing major stages of earth's geological history; be outstanding examples of significant ongoing ecological and biological processes; and/or contain the most important and significant natural habitats for conservation of biological diversity, including those containing threatened species of outstanding universal value from the point of view of science or conservation.[8]

The new reef park authority took the lead on the submission, showing that not only did the Great Barrier Reef meet all four criteria, but it also demonstrated significant cultural heritage value as well, embodied in its numerous shipwrecks and historic lighthouses, and places of deep Indigenous cultural significance.[9] Unsurprisingly, the application was successful, and on 26 October 1981, the Great Barrier Reef was officially listed as a World Heritage Area, recognised as a place of outstanding universal value.

In its determination, UNESCO described the GBR as 'the world's most extensive coral reef ecosystem', whose land and seascapes provided 'some of the most spectacular maritime scenery in the world'. It went on describe the reef's huge numbers of species and interconnected habitats, making it 'one of the richest and most complex natural ecosystems on earth'. 'No other World Heritage property contains such biodiversity', which means that the GBR is 'of enormous scientific and intrinsic importance' containing 'a significant number of threatened species'. The IUCN evaluation stated that 'if only one coral reef site in the world were to be chosen for the World Heritage List, the Great Barrier Reef is the site to be chosen'.

The great reef protection quartet had been accomplished. Preserved as a national park, managed by a Commonwealth statutory authority, backed up by a dedicated science agency, and recognised by the United Nations as a place of outstanding universal value. Surely now, the reef was home and safe. And yet, even seemingly well-established institutions are vulnerable to the kind of politically motivated misinformation and unfounded accusation that was beginning to rise up around us.

12

Ambush

If corals are exposed to consistently elevated temperatures, they will eventually expel the algal symbionts which give them their colour, and all that remains is the pure white calcium carbonate skeleton. The corals have 'bleached', and without their symbionts they will soon die. But if temperatures cool quickly enough, the symbionts can return, and the corals can recover.

As Covid gathered pace in late 2019, we watched anxiously as sea surface temperatures on the reef rose. Long, hot, cloudless days stretched one after the other, and heat started to accumulate in the reef lagoon waters. By early 2020 the first indications of bleaching were becoming apparent, and by March it was clear that the GBR was experiencing another mass bleaching event. We watched in dismay as the events of 2016 and 2017 appeared to be repeating themselves. And then, suddenly, the weather changed. A front moved in. Cloud cover built, shading the reef. Rain came, cooling the land and the water. And, just in time, sea temperatures dropped. By midyear, it was clear from our monitoring that most of the bleaching we'd observed had reversed. The symbionts had returned and the corals soon regained their colour and vibrancy. The cooling trend had arrived just in time. That summer of 2020, the reef had got lucky.

We in the scientific community heaved a collective sigh. It could have been so much worse. If above-normal temperatures had continued for even a week or two longer, a lot of coral would have died. But there was no time for reflection. Even with Australia isolated and the world shuddering under the worst pandemic since the Spanish Influenza a century before, the effort to discredit reef science was building.

The Senate inquiry into the effects of farm practices on reef water quality that kicked off in 2019 had been delayed by Covid, but submissions were now in, and on 27 July 2020, public hearings began in Brisbane. Senators Green, McDonald, Rennick, Roberts, Sterle and Waters were in attendance, but due to the Covid travel bans affecting much of the country, many attendees, including Britta Schaffelke and I, joined by telephone. That, as we were soon to find out, was a mistake.

The hearing started out conventionally enough. The committee chair, Senator Sterle, welcomed the participants and witnesses, and reminded them that 'in giving evidence they are protected by parliamentary privilege. It is unlawful for anyone to threaten or disadvantage a witness on account of evidence given to a committee, and such action may be treated by the Senate as a contempt. It is also a contempt to give false or misleading evidence to a committee.'

The first witness was called, Jane Waterhouse from James Cook University, an acknowledged expert in marine water quality and catchment processes, and one of the authors of the 2017 Scientific Consensus Statement on water quality. Her opening statement encapsulated the existing evidence. Declining water quality associated with land-based runoff was one of the key drivers of the current poor state of the Great Barrier Reef. This connection between declining water quality and reef health was not a new one, she pointed out, nor was it 'unique to the Great Barrier Reef'. The foundational evidence for this conclusion was 'very strong and well-established'.[1]

She continued: 'There are multiple lines of evidence that have identified increased loads of sediments, nutrients and pesticides to the Great Barrier Reef since European settlement. There is little dispute whether or not these loads have increased. There are a range of methods that enable us to make those conclusions ... and the science is continually improving.'

As Britta and I listened, she went on to lay out basic findings of the Consensus Statement. Sediment from runoff affected near-shore reefs most, particularly after flood events. There was evidence of pesticides making their way to the GBR, and 'the evidence that it comes from agricultural industries is unequivocal'. Similar effects had been observed around the world, wherever ecosystems were downstream of agriculture. Finally, she stressed that water quality influence on the reef was 'largely within the high-value coastal and inshore marine ecosystems', the parts of the reef most highly valued for recreation and tourism. The reef, she reminded senators, was not 'just about coral'. It was a complex ecosystem of interconnected communities, including 'intertidal areas, seagrass, mangroves and fish communities', all of which were linked to the land catchments.

It was an eloquent and powerful beginning. Britta and I were next up, buoyed by our colleague's strong performance. I was invited to make an opening statement. After briefly summarising AIMS's role and current standing (ranked first in marine science globally), I summarised the key points of our written submission, namely that our long-term monitoring of the reef showed that it was 'in a period of prolonged decline due to the combined impacts of deteriorating water quality, cyclones, crown-of-thorns starfish outbreaks and the major threat, climate change'. AIMS research showed unequivocally that deteriorating water quality had a negative impact on coral reefs and other coastal marine ecosystems, and that despite years of effort, inshore water quality on the GBR had not improved overall, and in many places had worsened.

I concluded: 'The extent of this impact is mostly constrained to coastal and inshore marine systems. Because these areas are comparatively easy to access, they are of especially high economic and social value to recreational users of all kinds, the tourism industry and traditional owners. As time goes by, disturbances are occurring more often, are longer lasting and are more severe. We now know that chronic human impacts such as poor water quality, especially in near-shore areas, are superimposed on more acute long-term climate-related pressures, making it harder for ecosystems to recover. The current trajectory of reef health is unsustainable. If the reef as we know it is to survive the coming decades, each of the various sources of stress that we can control will have to be reduced. The key question is how. It's time for new approaches and new solutions. Clearly, business as usual is not working.'

At this point the chair opened the floor to questions.

Susan McDonald, a Liberal National Party senator from Queensland, was first to speak. Highly intelligent and personable, McDonald comes across as genial and supremely competent. In previous interactions, we had always got on well.

'I want to start with Dr Hardisty,' she said. 'I understand that AIMS is concerned about coral growth rates and that coral calcification rates are an important measure for the health of coral. That's why I was so pleased to read this in a letter that your organisation wrote to AgForce on 29 May: "Coral calcification rates in *Porites* are not significantly impacted by agricultural activities." For those who are not aware, *Porites* are the species of large corals that are used to measure growth rate data from. This seems to be great news. Farming is not significantly impacting coral growth rates in *Porites*, as you've said. Why is this not something that I've been able to read in any of the consensus documents?'

It was not the opening question we had expected. Suddenly, it was clear what direction the opponents of Queensland's new reef regulations were going to take. Somehow, Peter Ridd's coral growth rate obsession had become their frontline issue.

I responded. *Porites*, the massive, slow-growing brain corals, were among the hardiest and longest-living corals on the planet. This made them ideal for coring, providing us with an ability to look back in time. But they were only *one* of hundreds of species of coral present on the reef. All the coral core work that we'd done was looking at the influence of temperature on declining coral growth rates, and that link was clear. Bleaching due to increased temperatures definitely slowed growth rates. But 'we have never linked declining coral core growth rates in *Porites* – that big brain coral – to runoff,' I said. 'They're really two separate issues that I think, due to some direction from outside, were being conflated.'[2]

Britta jumped in to support me. This issue of coral calcification rates in *Porites* was 'not something that has anything, as far as we know at the moment, to do with water quality, and that's why in the most recent Consensus Statement, which was an update of the state of knowledge, it was not specifically mentioned. The work, firstly, has been done in the past, and it also wasn't entirely relevant.'

After several additional questions about the effects of pesticides on coral growth rates, McDonald's questions shifted to our involvement with the new regulations and the quality of our science. Was AIMS doing research that nobody knew about or could access, particularly the people who were most affected, and what input had we provided to these regulations?[3]

I replied that we only provided the science. We didn't get involved in how that science was used develop policy. We remained independent and objective.

We were trying to answer the questions, but it seemed to me that our answers were beside the point. A record was being created; an initial position established. As Senator McDonald moved on to 'this issue of quality assurance of the science', I wanted to ask what she meant by *this* issue? It could mean only one thing. I could now see very clearly how the rest of day was going to unfold.

'I notice that AIMS has made a point now that it has a red-blue team approach* and that you have implemented this quite recently,' she continued. 'Were red and blue teams implemented to improve quality assurance?'

'Yes, that's exactly what we're trying to do. As I said in my opening statement, we're striving for continuous improvement. We want to make sure that, every year that goes by and every decade that goes by, the science we do is better than it was ten years ago. Technology improves. We learn more. We gain insight. That's what science is about. It's a journey of building from knowing nothing two hundred years ago to knowing more and more and more. So that's how the process works.'

'So can we assume that the earlier work done at AIMS prior to 2017 was not as rigorously checked because it did not utilise this approach?' *Gotcha.*

I steadied myself, left the mute button on for a moment before replying. 'What you can assume is that this is one more approach that

---

* The red-blue team concept originated in the United States, where it was used to simulate military attacks on the United States and its interests. The red team acts as the attacker, and attempts to defeat the blue team, charged with defending the country. The idea has been widely applied since in areas such as cyber security and law enforcement. AIMS started using it in 2018 to test our own work and conclusions, adding to the internal and peer-review quality control protocols already in place.

we added a couple of years ago just to try to make it even more airtight. I don't think it's that it was bad before; it's that it's getting better. A mobile phone from twenty years ago worked okay: you could have a conversation. But smartphones are better now. Why? Because you have twenty years of development and improvement under your belt.'

'Does AIMS believe that other scientific institutions could benefit from the use of similar systems like the red-blue approach?'

'Yes.'

'Are they used in all other institutions?'

'I couldn't tell you.'

'Does AIMS, then, support the idea of a formal red-blue team organisation that would help with consensus statements and other major synthesis documentations?'

'AIMS supports anything that will increase and improve our knowledge over time.'

The scene was now set. And some of the other committee members could see it too. Labor's Senator Green asked me 'whether you see any threats that exist in discrediting the science and research of your organisations. What is the effect of campaigns around discrediting that science?'[4]

I replied, clumsily, that you wouldn't go to a plumber to fix your car, you'd go to the experts. And we and our colleagues were the experts. 'Nobody's saying that everything that has been done in the last hundred years ... has always been correct,' I went on. Science advanced by testing hypotheses, building on knowledge, and learning from mistakes. 'I think everybody who's done basic science at high school knows that. Using ... the process of advancement as a whip against the eventual results that are honed and developed and tested over decades – yes, I think that's pretty cynical. But we live in a technological and scientific society, don't we?

Look all around us. Look at everything that we rely on. To the degree that those things are discredited, I think we're poorer for it.'

If we had hoped that this change of direction would move the inquiry in a different direction, we were wrong. Senator Rennick, next to question, brought us swiftly back to where we had started.

'My question is to AIMS,' he began. 'I understand that there is some data on coral growth rate since 2005, but there is no recorded publication that gives a Great Barrier Reef-wide average of growth data since 2005. Is that correct?'

Britta answered. 'The piece of evidence you are referring to here is the analysis of calcification rates in coral, of *Porites* coral, this long-living tolerant coral—'

'No, no,' interrupted Rennick. 'This is a different question.'

Undaunted, Britta continued, 'Where a publication in 2009—' but was immediately cut short again.

'Excuse me, no. You're not answering my question. Is there an overall set of data that shows the growth rates or declining rates of coral growth across the entire Great Barrier Reef? Yes or no.'

'With all due respect, I have been answering your question. So—'

'Just a yes or no, please,' Rennick directed, his voice terse.

'Don't harass her,' interjected one of the other committee members. Listening on the phone it was hard to tell what was going on, and who was speaking.

Rennick pushed ahead. 'No, I want the answer to the question. It's very simple. Is there an overall database of coral growth rates or declines for the entire Great Barrier Reef basin?'

I jumped in. 'I think I've got to point out that we've already told you that this has got nothing to do with this inquiry. Coral growth rates—'

Again, I was cut off.

'So you can't answer the question, is that it?'

'We have never connected them to anything to do with farm runoff or farm activity—' but once more I was interrupted before I could finish. At this point, I was fuming. Joining only by telephone, I couldn't read the body language of the participants or bring any physical communication to bear. What was going on felt to us like straight-out harassment – this was not a court of law, and we were not witnesses in a murder trial, unless you considered what was going on here to be another stab in the murder of the reef.

This was my first encounter with Senator Rennick. Subsequently I was to learn that he considered climate change to be 'junk science' and claimed that greenhouse gases don't heat up the earth.[5] This despite the fact that the link between CO_2 and atmospheric heat accumulation was scientifically proven by Arrhenius in 1896.[6]

At this point, proceedings deteriorated into farce. Every time we attempted to answer Rennick's questions he talked right over us, hammering on about the same question.

'I just want to know if the coral growth in the Great Barrier Reef increased since 2005 or decreased.' He was clearly as exasperated as I was. 'Is there a database? Is there somewhere I can – can you give me a link to the data, comprehensive data, of the entire Great Barrier Reef, of the growth rates, since 2005, or the decline rates?'

I kept trying. 'The question makes no sense, sir, with respect, because the Great Barrier Reef—'

'Well, excuse me—' Rennick said, not letting me finish.

By this point, I'd had enough of being talked over. I kept going: '—in its entirety is not made up of *Porites* coral.'

'I'm talking about all coral.'

'It's just one species of many, many corals—'

'Exactly.'

'So an entire database of growth rates for the Great Barrier Reef as a whole? No such thing exists.'

And then, suddenly, Rennick changed tone. 'Thank you,' he said. 'That's all I wanted to know. So my question is: if we've been spending billions of dollars on doing research into the Great Barrier Reef and coral in the Great Barrier Reef, why haven't we got one comprehensive knowledge management system that shows what the growth rates of coral, of the various rates of coral, along the various reefs? ... How are we supposed to make legislation if we can't tell what's going on? I've spent the last number of days going through all these submissions and I see nowhere any benchmarks that show me what declining water quality means or what the rate of coral was in the 1880s, what it was in 1970, what it was in 2010. So I've got no idea what to believe here ... I've got issues with the credibility of our research.'

But we were not given the chance to respond to any of this. Without pausing, Rennick moved directly into his next question. 'Have you had any input into the best practice management plans implemented by the state Labor government?'

Britta tried to go back to Rennick's issues about the credibility of our science. 'I would like to have one minute ... to ponder that question, because, if you are interested in the role of—'

But he wasn't interested, and showed it by again talking right over Britta. 'No, could you answer this question now. I've moved on.'

Britta kept going. '—coral on the Great Barrier Reef, that is data—'

'Can you answer my question? Have you had any input into the best practice management plans implemented by the Queensland state Labor government?'

At this point, I couldn't help myself. 'If you want an answer to these questions, accept that when Britta is speaking, she's going to get there and answer your questions. I'm just on the end of a phone line, but I hear voices overlapping—'

Finally, the chair intervened and gave Britta the chance to continue.

'Thank you very much,' she began. 'I just wanted to add that there is a complexity between growth rates ... We have a 35-year dataset that shows the coral cover growth and decline for the whole Great Barrier Reef, so that is available. That is published every year as an update from the long-term monitoring program. This is the growth of coral communities, which is much more relevant. It's probably the information that Senator Rennick was looking for, but he referred to a specific case of the *Porites* corals, which we have discussed earlier, which—'

But once again Rennick wouldn't let her finish. 'No,' he clipped, 'I didn't refer to the specific rates. Would you not put words in my mouth?'

'But, yes, there is data there on the state of the reef—'

'Would you not put words in my mouth? And you've just contradicted the previous guy who said it's not available, so now I'm confused. Anyway, we're moving on from that. I'm taking it as not available. Next question: have you had any input into the best practice management plan implemented by the Queensland state Labor government? Yes or no?'

'Senator Rennick, I answered that. I said that, no, we didn't—'

Another talk-over drowned Britta out, and this time Rennick did not stop.

Finally, the chair interjected. 'Order!' he shouted, the sound coming through as a screeching distortion over the phone. 'Senator Rennick, if you have frustrations because your questions aren't being answered directly – and I can relate to that – I urge you, while we are on show in your great state, to target your questions to the terms of reference that we have in front of us. Senator Rennick, do you have any further questions?'

He didn't. But it wasn't over yet.

The chairperson recognised One Nation's Malcolm Roberts, the same glacier-eyed senator who had launched that attack on reef science in the Senate chamber back in 2016. Roberts wasted no time bringing us back full circle again.

'My question is to AIMS to start with. Let me just understand, and I want to confirm, that AIMS has not published results of what the Great Barrier Reef average coral growth rate has done in the last fifteen years. Yes or no?'

'Yes, we have published extensively on the state of the coral reef in the last period that you have identified.'

'That was not my question. I'll ask it again: AIMS has not published results of what the Great Barrier Reef average coral growth has done in the last fifteen years – the rate of coral growth. You published it from 1990 to 2005, claiming a collapse in growth rates. What have you done since 2005 to now on growth rates?'

'Growth rates in *Porites* coral – is that what you're asking?'

'Growth rate across the Great Barrier Reef akin to what you published and what was quoted from 1990 to 2005. Across the Great Barrier Reef, what has been published on growth rates for coral since then by AIMS?'

Britta now: 'I think I answered that before when answering Senator McDonald's question. We have not, since 2005, looked at a whole GBR assessment of *Porites* coral calcification rates, which is what you're referring to. We have published work that showed that the calcification rates have declined in response to thermal events – to bleaching – and then have recovered again. But we are not doing this assessment as a routine monitoring product.'

'That seems strange to me, because the eminent scientist Dr Ridd plus Dr Stieglitz and Dr Da Silva hotly dispute your claim. You had a claimed collapse in growth rate from 1990 to 2005, but nothing since – nothing – across the Great Barrier Reef.'

'There were publications, and, with all due respect, Senator Roberts, they were actually made available to the [people] that you mentioned, and they choose to not mention that newer work.' Britta's voice cracked with frustration.

Roberts followed the same pattern as Rennick, switching now to the red and blue teams question. Once again, the suggestion was that because we had recently added a new quality control measure for some of our work at AIMS, then everything we and others had done before – including the 2017 Consensus Statement on water quality – was necessarily suspect.

As if following the script from one of Ridd's lectures to farmers, Roberts now switched to the 'replication crisis'. Building up the questions as would a cross-examining barrister, he tried to manoeuvre the witness – me – into a corner.

'Do you also agree that replication and repeating experiments by scientists is crucial to the scientific method?'

'Yes.'

'Are you aware of the replication crisis, where it is regularly found that a large fraction of the peer reviewed literature – maybe 50 per cent in some estimates around the world – is in error.'

'Absolutely not.'

'I understand that Dr Ridd has asked for a replication study of the AIMS coral growth rate that supposedly shows that coral growth has drastically slowed between 1990 and 2005. Given the importance of that data and the claim of decline in the coral, do you support a study by other scientists to replicate your data to make sure it is all okay?'

'With respect, two points: as I've mentioned a couple of times before, in terms of this inquiry the coral growth rates of *Porites*, which you're talking about, we have never connected it with anything to do with farm runoff. So, in terms of the chair's guidance to stick to the terms of the inquiry, I don't really see what the point is.' I laughed as I said it.

'Well, I take your laughter.'

'However, if you want to get into growth rates of *Porites*, that's a whole other story. As we've said, it's not connected to farm practices – we don't believe it is.'

'I'll ask again: Dr Ridd, who's an eminent scientist, has asked for a replication study of the AIMS coral growth rate data that supposedly shows that coral growth drastically slowed between 1990 and 2005. Would you support a replication study – another study to replicate that?'

'If somebody wants to give us the amount of money that's required – it's very, very expensive work to redo it – then, yes, I'm always happy to redo a study. I don't think it would be particularly useful, though, because we have already supplemented that work, as we have repeatedly written to and communicated with Dr Ridd about, over and over again. As my colleague Dr Schaffelke indicated, he doesn't seem to want to listen to any of the new information we give him. He's got a monotonic view on this. No matter what information or additional papers and studies we provide him with, he seems to ignore it all. I don't really have anything else to say about that.'

'I didn't ask for comments about Dr Ridd; I asked for comments about your studies and whether they should be replicated.'

'Well, I'm just answering the question. You keep mentioning Dr Ridd – eminent scientist Dr Ridd.'

Roberts pressed on with our JCU colleagues for a while, following the now-familiar lines. Asserting again that the peer review process was suspect, that major papers associated with climate change science had never been checked, plying the same message of doubt in science in general, and reef science in particular. It was like being stuck in an episode of *The X-Files*: 'Trust no one.'

After a few questions from senators Canavan and Waters that sought to elicit information rather than to score points, a break was called and we were released. It had been just under two hours, but it had felt like a very long day. I remember standing in my office after disconnecting from the hearing, looking out of the window, thinking: this is not over. Not even close. In the following days, our testimony would be twisted, misrepresented and weaponised against other witnesses, sparking formal complaints and a motion of censure in the Senate.

~

After we had been dismissed, the hearings continued. Senators heard from witnesses from various cane farming groups. Paul Schembri, chairman of Canegrowers Queensland, spoke eloquently about the progress made by farmers in meeting best management practices designed to reduce flows of sediment, nutrients, and herbicides and other chemicals to the reef. He also pointed out the significant economic contribution that the industry made to Queensland, and the cost to farmers of the increasing regulatory burden. The general manager of the Australian Cane Farmers Association, Stephen Ryan, made the sensible suggestion that more consultation was needed between government, researchers and farmers.

But other representatives from the industry continued in a vein reminiscent of the Roberts-Rennick line. Michael Kern of the Burdekin District Cane Growers took up the charge. 'Cane farmers do take their environmental stewardship responsibilities, including the health of the GBR, very seriously,' he testified. 'They've trusted reef scientists to get the science right and tell the real stories about what's going on at our 3000 or so pristine reefs. That trust has been destroyed. Instead, cane farmers are being publicly demonised and may soon wear penalties of over $200,000 for a crime they did not commit – and, more significantly, for a crime that never happened – using unfair laws based on unchecked science and unproven modelling. Our evidence tells a different story. The real GBR is not dying. Farming runoff doesn't even make it out to the real Great Barrier Reef. Water coming onto farmers' farms is being filtered, thereby improving the quality of water passing through their farm. The integrity, certainty and reliability of the reef science has been further eroded by a couple of recent events – firstly, some of the admissions by AIMS. We heard one challenge today in terms of the coral calcification rates being a stress response to extreme heat events and not to the effects of farming … Secondly, the currently available data for the growth records derived from coral cores are not fit for purpose to differentiate between growth rates for areas close to agriculture and non-agricultural areas.'

Admissions? We had provided information to the committee, nothing more, nothing less. And here that information was not only being misrepresented, but being made to appear as if it had been pried out of us under withering cross-examination. This theme would be repeated throughout the rest of the day.

Julie Artiach, also of Burdekin District Cane Growers, was even more blunt. 'The Australian Bureau of Statistics reported that for the 2019 financial year the Commonwealth, state and territory governments spent $3.33 billion of public money on R&D, the by-product of which often formulates government policy. The example before us today is the 2017 Scientific Consensus Statement and the basis of that statement supporting the 2019 amendments to the Queensland *Environmental Protection Act*. There appears to be consensus in the scientific world, regardless, I think, of the comment made by the scientist from AIMS this morning that there is a replication crisis in science.' AIMS had said no such thing.

She continued: 'In other words, and under a more stark reality, this is research fraud. Therefore, it mandates that, as a matter of evidence-based policymaking, which is incumbent on government, this evidence must be robust and subject to oversight.'

Under questioning from Senator Waters, however, Dan Galligan of Canegrowers made clear the core of the issue. 'We absolutely agree that farmers have lost all confidence in the science, though, and that's because of *the way the science has been used to make a decision* [my italics]. It's the State Government that made the decision to regulate, and they used science to make that decision, and I think they used it badly ... But the way in which the government has used science and interpolated science to make a decision to regulate is what we're actually discussing here, and I think it's been a very bad outcome. It's been a very bad outcome for science and it's been a very bad outcome for our industry.'[7]

This was pretty much what I had agreed with Paolo during my walk around the farms in his area. In their fury at the Queensland Government's new regulations and the lack of consultation in their formulation, some farming groups had latched onto the broader attack on reef science being driven by climate denialism as another way of making their point. How ironic, then, that cane farming is one of the most carbon-friendly industries in Australia, having operated at very nearly close to net zero emissions for many years now. It is something they quite rightly celebrate on their websites. Another irony was that for all the brouhaha about coral growth rates, we had stated very clearly that declines in calcification rates were due *not* to farming runoff but to warming ocean temperatures, and that while water quality was a concern for the reef, climate change was by far the most serious threat to its future. I could reach only one conclusion. In their time of need, upset and vulnerable, farmers were being used by the climate denialist lobby to amplify the message of distrust in science.

Despite this, Rennick and Roberts kept hammering away with their yes-or-no questions (which allow for no scientific explanation), twisting and misrepresenting our testimony from earlier in the day, which created a cloud of confusion and mistrust. The regulations were based on 'a concocted science that has been fabricated against your business'. The science was 'distorted', and regulations were 'concocted'.[8]

Stephen Lowe, chair of the Australian Banana Growers Council provided a powerful voice of reason. He believed that 'there needs to be greater respect and recognition of the efforts growers are already making to reduce runoff from their farms ... [Growers] are making significant investments and are choosing significant changes to the way they farm their land. Applied nutrient levels are decreasing and areas of vegetation coverage are increasing. Yet these changes are not showing up in the modelling and the report card. There is a disconnect that is causing resentment and scepticism. Growers are questioning the science because the science and modelling are not reflecting their reality. Government researchers and industry need to work together to agree on a way to measure and show the progress that is actually happening. In my opinion, the current method of reporting is broken. Water quality science needs to be demystified for growers. More collaborative research between industry, community and government is currently happening in Wet Tropics Major Integrated Projects, with growers in the Johnstone and Tully catchments involved in designing trials, taking water quality samples, and learning about the link between spraying practices and risks to water quality. This only occurs because the information is shared in a way growers can understand. I'm not convinced that a new office of science quality assurance would be the silver bullet to improving the robustness of water quality science. I fear it may add a new level of bureaucracy and more confusion.'

But his was a lone voice. Spurred on by Roberts and Rennick, the story became increasingly distorted as the day went on, becoming, at one stage, a circular argument of bewildering simplicity, a breathtaking logical fallacy, grasped and repeated by other witnesses. AIMS's work on growth rates in *one species of coral*, the hardy long-lived *Porites*, was suddenly representative of *all* corals, despite our repeated assertions to the contrary. And since we had testified that decline in *Porites* was not the result of farm activities but was tied to ocean warming and climate change, it was deduced that *no* corals were affected by deteriorating water quality from farm runoff. It was more than frustrating.

In Rennick's own words, in a question to CSIRO later in the afternoon: 'This morning the Australian Institute of Marine Science and TropWATER said "it was never the farmers' fault", "it is not connected to farmers' practices" and "can't give specific impact on coral growth rates by herbicides or pesticides". The entire panoply of issues challenging the Great Barrier Reef had been reduced to the single pronoun 'it'.

And despite evidence from numerous scientists that water quality *was* affecting the reef area, not only corals but other habitats such as seagrass meadows, the reef was regularly described by those pushing the untrustworthy science line as 'pristine', 'fabulous', and of course 'untouched', as if by sheer repetition these assertions could be made true for every part of the reef.

Peter Ridd was next to appear. In a session that sounded more like a scripted fireside chat, the eminent scientist and the One Nation senator chatted about the definition of science, the need for quality control and the imperative of challenging scientific results – stuff that any scientist will agree on. But their conversation quickly descended into the bizarre.

'I have a series of yes/no questions,' stated Senator Roberts. 'In these days of policy-driven pseudoscience, politicians and journalists – and, sadly, many academics – use alternative science that falsely masquerades as science. These include claims of consensus. Is a claim of consensus a replacement for science?'

'No, it's certainly not,' replied Ridd. 'If we always went with the consensus, we would never get anywhere in science.'

'Is the use of emotion – fear of catastrophic consequences, guilt, pity – a replacement for science?'

'No.'

'Is the use of "if" statements – if Antarctica melted, the world would be flooded – a replacement for science?'

'Not in the sense you mean, no.'

'Is peer review the way it's conducted loosely these days – it's really buddy review – a replacement for science?'

'No. It's a useful system as a first go, but it doesn't replace it, and it's not a quality assurance system to be used to place regulations on a large number of farmers.'

'Are celebrities, actors and socially awkward sixteen-year-olds a replacement for science?'

'No.'

'Are smears and ridicules, including derogatory labels like "deniers", "sceptics" and "conspiracy theorists", replacements for science?'

'No.'

'Are lies, data distortion and cherry-picking replacements for science?'

'No.'

'Are claims about donors, whether false or real, replacements for science?'

'No, they're not. It is relevant information to know where the money has come from, but they're not replacements. You've got to look at the data; it doesn't matter who paid for it. They're interesting, but they're not replacements.'

'Is emotively claiming that natural legacies like the Great Barrier Reef, Bondi Beach or Kakadu are threatened replacements for science?'

'It's not a replacement, but it is important in our consideration.'

'Is indoctrinating kids a replacement for science?'

'No.'[9]

The senator's penchant for monosyllabic answers had been rewarded, and as a bonus he'd managed to work in a thinly veiled disparagement of Greta Thunberg, the teenage climate activist. The Karl Rove playbook was again in full evidence. And in many cases, it appeared to be working. Several witnesses from farm organisations mentioned seeing Ridd's talk, and how that got them to question the science.[10] Others quoted his positions almost verbatim.

The second day of the hearing continued pretty much in the same vein. Senator Roberts kept disparaging science with undulled enthusiasm, calling Ove Hoegh-Guldberg's work for the UN's IPCC 'climate fabrications', and enquiring of Richard Leck of the WWF conservation group if Professor Hoegh-Guldberg was funded by them.[11]

Wait a minute, what about yesterday's 'are claims about donors, whether false or real, replacements for science' question? Never mind, it didn't matter.

Senator Rennick continued to harangue witnesses and misrepresent AIMS's testimony, to the point where one witness, Dr David Wachenfeld, chief scientist for the GBRMPA, responded: 'I was listening to the AIMS testimony. I'm sorry to have to say that I have a different interpretation of what they said.'

Things got so bad during questioning of witnesses from Divers for

Reef Conservation that Senator Waters felt compelled to apologise to the witnesses 'for the line of questioning that you've just been subjected to'. When Rennick attempted to interject, Waters immediately countered: 'No, you don't get to interrupt. You were rude to the witness, and I'm apologising on your behalf and on behalf of the Senate for bringing us all into disrepute.'

There would be one more day of formal public hearings later on in August, and then the committee would table its final report in October.

In the aftermath, AIMS was compelled to lodge a formal supplementary submission, decrying the misrepresentation of our testimony, and correcting the record. Others, too, appalled by the treatment received by witnesses appearing to present the scientific case, wrote to the committee in protest. Among them were Professor Ian Chubb, chairman of the Independent Expert Panel advising the minister on the reef, Professor Ove Hoegh-Guldberg, and the president of the Australian Academy of Sciences.[12]

The final report was based on 120 formal submissions and the testimony of dozens of witnesses. It opened with an unusual statement:

> During the course of the inquiry, concerns were shared by science representatives about the conduct of some senators during the public hearings. In correspondence from Professor Ian Chubb, Professor Ove Hoegh-Guldberg and Dr Geoff Garrett, the committee was advised that their participation in the inquiry was in the belief that the inquiry was a 'genuine attempt to understand better the extraordinary and interrelated complexity of the [Reef's] ecosystem'.

It further quoted the scientists' statement:

Given the disturbing nature and tone of the hearings in Brisbane in July, we should have anticipated the discouragement we now feel. We should have sensed that when some Senators don't want to know about the seriousness of an issue, they tune out anything that is inconsistent with their fixed views and preconceived ideas. And no matter how much evidence they are given, or its rigour, nothing changes if it doesn't fit what they want to believe.[13]

The report stated that

AIMS were forced not once, but twice, to write to the Committee to raise concerns that its evidence was being misinterpreted. Its second correspondence noted: 'Despite our previous supplementary submission to this enquiry, which corrects numerous misrepresentations of AIMS' testimony ... we now find that such misrepresentations have continued in subsequent sessions.'[14]

The report also quoted Ian Chubb, Dr Geoff Garrett, former chief scientist for Queensland, and Ove Hoegh-Guldberg, who described their experience of appearing before the committee in the *Guardian*: 'We saw witnesses talked over and have their longstanding commitment to their scientific field transparently treated with contempt. All because their evidence didn't fit the apparently pre-conceived and intransigent views of some of the Senators participating in this inquiry.'[15] It was a damning rebuke.

In the end, however, after all the acrimony was stripped away, the report provided a sound and comprehensive summary of the challenges facing the reef, the underlying state of scientific knowledge, and made a number of useful recommendations. The committee backed the findings of the 2017 Consensus Statement, and the robustness of the process it used to reach its conclusions.[16] However, the committee did express concern about the 'misunderstandings about the Consensus Statement by those stakeholders impacted by its findings'. On this basis the committee's first of seven recommendations was that 'the Australian and Queensland governments ensure adequate stakeholder engagement and education processes are integrated into future Water Quality Scientific Consensus Statement processes'.

Despite the claims by some committee members, the report strongly backed the science, also pointing out that while scientific knowledge of the reef was clearly not complete, and likely never would be, it noted that the Australian Government, under the *Great Barrier Reef Marine Park Act* 1975, is governed by the precautionary principle, which means the 'lack of full scientific certainty should not be used as a reason for postponing a measure to prevent degradation of the environment where there are threats of serious or irreversible environmental damage'.[17]

The committee stated that it was 'extremely concerned by the unsubstantiated claims made by various witnesses and submitters about the adequacy of the peer review process'. It did not share the view that

> the peer review process is inadequate or lacking, that it produces group think or corrupt behaviour, nor that it has created a replicability crisis within Reef science. Further, the committee does not believe that Australia's scientific community is intentionally covering up flaws in research and vilifying or excluding those scientists that raise concerns with the quality of the research. Rather, the committee is reassured by those scientific representatives that spoke of the sector's incorporation of dissenting and alternative views, and their encouragement for those scientists to submit their work through the peer review process.[18]

The committee concluded by rejecting calls for Ridd's so-called 'Office of Science Quality Assurance'.

With respect to the fundamental issue of the inquiry – farm practices and their effect on the reef – the committee's view was conciliatory. It congratulated the agricultural sector for its widespread adoption of best management practices, but remained 'concerned by ongoing reports of poor water quality in the Reef, and the slowing down of [Best Management Practice] uptake by the sugarcane and grazier sectors'. On current trajectories, the Reef 2050 Plan water quality improvement targets for nutrients 'would not be met until 2035, a decade after the agreed target date'. The committee recognised that 'the Queensland Government's regulations are necessary to expedite rapid uptake of best practice', and it identified that 'further work should be done by the Australian and Queensland governments to ensure consistency in reporting for the agricultural industry best practice'.

The committee also recognised that the new regulations imposed additional costs on the agricultural sector (by their estimate about $250 million over ten years), but that implementation of best practice could also save them money (about $285 million over ten years).[19]

In the end, three senators put their names to a dissenting report which appeared as an addendum to the main document. While agreeing with many of the recommendations of the main report, they also called for the establishment of an office of science review, a reconsideration of the Reef 2050 Plan water quality targets, a rollback on penalties to farmers, and a review of the applications of the precautionary principle in reef management.[20]

Another protracted skirmish was over, with no clear wins for anyone. Farmers remained concerned about the financial burdens being placed on them, and the lack of consultation on matters directly affecting their livelihoods. Scientists recognised that they needed to do more to engage directly with the agricultural sector to explain the complexities of the reef – and saw clearly that if they didn't, misinformation would fill the vacuum. Governments realised they had a lot more to do to align and unite stakeholders, particularly as emotions ran high. And the politics of the reef had just become nastier, more divisive and more personal than ever.

13

Warnings, Progress and Setbacks

The past two hundred years has shown that the human spirit of discovery, activated and enabled by the formal process we now know as science, will not be denied. Whatever is going on in the world, scientists will continue to delve into the mysteries of our universe and seek the answers to humanity's challenges.

Despite the turmoil around the Senate water quality inquiry and the concerted attempts to discredit science, scientists went about their work. Data was collected. Observations made. Theories postulated and tested, conclusions reached and possible solutions devised. The business case for the Reef Restoration and Adaptation Program (RRAP) had been externally peer reviewed by two highly respected, independent international scientists, and then endorsed by Professor Chubb's Independent Expert Panel. Funding had been agreed, and money had started to flow from the Great Barrier Reef Foundation's now infamous $443 million grant from the Commonwealth Government. A formal governance structure for RRAP was established with Professor Rob Vertessy, ex-CEO of the Bureau of Meteorology, as independent chair.

An executive director was hired and AIMS appointed as the managing entity, with the job of bringing together a consortium of institutions including CSIRO, The University of Queensland, JCU and Queensland University of Technology. The GBR Marine Park Authority would act as an observer. After all the brouhaha, the research was about to get underway as originally envisaged.

The task at hand was monumental. With $100 million from the Commonwealth via the GBRF grant, and matching funds from the research providers themselves and from the foundation's own fundraising, RRAP had a notional $300 million to work with over five years. Sitting down with David Mead, the program director, it was pretty clear from the outset that even this wasn't going to be enough. The challenges were immense. Nothing of the kind had ever been attempted before, anywhere, and certainly not at this scale. We had to develop a set of techniques and processes that would allow us, if governments decided it was necessary, to intervene directly on the reef to improve its ability to survive marine heatwaves and adapt to those conditions in the longer term, and allow us to do it over hundreds and thousands of square kilometres in a way that was safe, affordable and acceptable to regulators, the public and traditional owners.

The enormity of what we were taking on had just started to dawn on us. We needed to test and evaluate of dozens of possible methods that might help the reef survive, from artificially brightening clouds to reflect back the sun and cool areas of reef suffering from the most intense heatwaves, to the breeding of heat-tolerant corals and introducing them onto the reef en masse. We needed to develop a new generation of reef models that would allow detailed prediction of the effects of interventions. We had to engineer solutions for scaling up prospective methods to each of the possible scales required. We needed to achieve widespread engagement with the public to understand and plan to overcome the social barriers to success. We needed detailed economic analysis of projected costs and benefits of various combinations of interventions over time, and an assessment of who would pay and who would gain. And we urgently required decision-making frameworks to guide the complex selection of what to do, where, when and to what degree.

David Mead had estimated that the entire research and development program would require ten years and perhaps double the amount of funding we currently had – or, as we were to realise, thought we had. Just bringing together the team – scientists, engineers, modellers, economists, social scientists and risk experts from a dozen different institutions – was proving a challenge. The reef community had never seen anything quite like it: project teams not assigned to individual institutions, but composed of the best people from among them; harmonised billing rates and structures; agreements to share intellectual property, and then make all new IP publicly available from the outset. It had been determined that this program was to be Australia's gift to the world.

There was so much to do. From the beginning, we had characterised this effort as a race against time. If we could find workable solutions and implement them before the reef had lost too much of its natural resilience, then maybe we could buy enough time for the reef while the world got its emissions reductions act together. Our initial modelling had suggested that a couple of decades, maybe three, might just be enough, if emissions could be brought down quickly. David and his start-up team were working sixteen-hour days to get things going, and you could see it in their faces.

And yet, we were making good progress. We had already done detailed evaluations of over forty possible intervention techniques, screened them down to less than a dozen realistic possibilities. The social scientists had completed an extensive public survey and determined that there was significant support for what RRAP was trying to accomplish. The teams were starting to come together well, driven by a strong sense of shared purpose and sure knowledge that time was short. Government had made it clear that while they understood that the full research and development program might take a decade, they wanted to be in a position to trial some of the most promising methods at pilot scale (a single reef covering a few square kilometres) sooner, perhaps within the next five or six years. It would be tight, but at this stage, we were confident we could do it, as long as the money was there.

But not everyone within the science community agreed that the RRAP research was worthwhile. Some detractors, such as Professor Terry Hughes from JCU – the scientist who had conducted the emergency aerial surveys of the reef to record the 2016 and 2017 bleaching events as they unfolded, one of the acknowledged world experts on coral reefs – had declined to participate in RRAP, and was now actively campaigning against the idea. His logic, set out in frequent social media postings, was simple. First, the reef, in Hughes's opinion, was simply too big for any human intervention to make a difference. And second, RRAP effectively gave Australia's conservative Liberal government an out on taking meaningful action on climate change. They could claim to be investing huge amounts in the reef, in this case to help it *adapt* to climate change, while doing nothing to actually solve the issue that was hurting the reef in the first place – namely carbon emissions. In cruder terms, RRAP was giving the government a 'get out of jail free' card. It was a compelling argument, and people were listening.

Whenever asked about this view, I simply replied that while I respected Hughes's opinion, we had to at least try. Global emissions of greenhouse gases had dropped by a few per cent during the Covid slowdown, but had already started growing again. In Australia the government was already talking about a 'gas-led recovery'. It was clear that emissions were not going to fall fast enough, no matter what happened over the next decade, to save the reef. Warming already locked into the climate system had contributed to the bleaching we'd seen that very year, a cooler La Niña year, and we had only just escaped widespread coral death. And we were still adding heat.

From now on, the reef was at risk of bleaching in any given year. I had said as much at Senate estimates. We could sit on our hands, hoping that emissions would somehow come down fast enough to make a difference, crying impotently as we watched the reef bleach, or we could do something now to help protect the reef in the short term. Yes, without strong cuts to carbon emissions, nothing we might do would make a difference. We had stated that repeatedly, in every RRAP presentation and report. Our modelling had shown that clearly. But if we could help the reef adapt enough in the short term – a couple of decades perhaps – it might buy the time we needed to bring emissions down.

But the Great Barrier Reef was no longer just ours. The world was watching, and the world was worried.

~

In fact, the world had been watching for a while. The years after the declaration of the GBR as a World Heritage Area saw an explosion of tourism. UN listing had greatly increased the reef's profile worldwide, and people wanted to see this unmatched jewel of the natural world for themselves.

Over the decade to 1990 the value of reef tourism increased sixfold. In 1987 tour operators alone took almost 1.5 million people to see the reef. By 1990 that number had doubled again, with another two million people visiting island resorts.[1] The pressure on the reef, especially in the heavily visited reefs off Cairns and in the Whitsunday Islands, was already resulting in serious damage, reminiscent of that seen at Yonge's Low Island after the publication of his book. Managing the new park was becoming increasingly difficult.

But during that period we were also learning a lot about the reef. The establishment of AIMS and the growth of marine science departments in a number of Australian universities led to a number of key scientific breakthroughs. A young Charlie Veron was busy compiling *Corals of the World*, the most extensive atlas of coral taxonomy ever assembled,[2] a work that would continue to occupy him for the rest of his life.

In the early 1980s the phenomenon of coral spawning was documented for the first time.[3] For a few days around the full moon in October and November, depending on the species, corals simultaneously release eggs and sperm into the water in a mass orgy of sexual reproduction that turns the ocean pink. Andrew Heyward was a young PhD student at the time, studying coral biology at JCU. He and a group of fellow students had started to question the conventional wisdom of the time, that all corals reproduced by internal fertilisation – the so-called 'brooders' – and then released larvae as fully-developed planulae that were ready to settle on the sea floor. 'Our perspective was biased by what we'd been taught,' he told me. 'We were conditioned to look for this behaviour.' But the theory seemed incomplete, so the students decided that closer observation was needed. 'We started to go out at night,' said Heyward. 'And that's when we saw it, this sort of snowstorm in the water as the corals released eggs and sperm. No one had looked at night before. It was like being in your garden and one night you see a flower bloom that you've never seen before, and then you realise that it only comes out at midnight, a week after the full moon. It was a really exciting thing to see.' Heyward and his colleagues quickly went on to identify the spawning behaviours and times of dozens of different species of coral. 'Now we know that about two-thirds of corals broadcast spawn like this, and only about a third are brooders.'[4]

The 1980s also saw mounting concern about the emerging issue of global warming. The second year of that decade, 1981, had been the warmest ever recorded. Reports from the Galapagos and Costa Rica in 1982–83 described exceptionally high water temperatures and severe coral mortality.[5] But the process was still a mystery. How and why were the corals dying? Nobody knew for sure.

The establishment of research stations on Lizard, One Tree and Orpheus islands gave scientists from across the country and around the world previously undreamed-of access to the reef. One of those scientists was the young Ove Hoegh-Guldberg, then a PhD student at UCLA in California. In 1987, he travelled to the Great Barrier Reef to study corals. His work, much of it conducted at the Lizard Island research station, identified for the first time the molecular process behind bleaching.[6] The mystery had been solved. Sustained exposure to elevated temperatures caused corals to expel their symbiont, without which they could not survive.

By 1987, as Ove Hoegh-Guldberg was starting his PhD, the world had set yet another temperature record.[7] That same year I was working for Shell Oil in Calgary. I walked into a bookshop downtown and happened to pick up a copy of *The Breathing Planet*. The book described the intricate workings of our planet's atmosphere, its evolution and the many threats it faced.[8] Leafing through the book, I came upon a graph from the Mauna Loa observatory in Hawaii that showed measured concentrations of CO_2 in the atmosphere. The line of data points saw-toothed its way in a steady climb from 317 parts per million in 1958 to 332 ppm in 1976. Each tooth of the line represented a single annual cycle of growth and dormancy in the northern hemisphere, plants taking up CO_2 in the summer for photosynthesis, and then releasing it during autumn and winter. It was as if the planet was literally breathing in and out. I hadn't yet read James Lovelock's *Gaia*, but the image of our only home as a single living, breathing entity hit me hard.

Part six of the book was called 'The Human Impact: the Carbon Dioxide Greenhouse Effect'. It described how mounting greenhouse gas concentrations in the atmosphere were warming up the planet, how we'd understood that link since the late 1890s. If we didn't change our ways, it concluded, the anthropogenic influence on the climate 'will probably begin to dominate over natural processes of climatic change before the turn of the century, and will result in a decided warming trend that will accelerate in the decades after that'.[9] Global average temperatures could rise by as much as 2 to 4 degrees Celsius by 2050, and the effects on weather patterns, rainfall distribution and ecosystems could be devastating.

I stood there in that bookshop reading until the sales clerk came over and asked me if I was going to buy something. I was enthralled, and scared. I still have the book. It changed my life. A couple of weeks after finishing it, I signed up to do a master's degree in environmental and water engineering at Imperial College in London, determined to spend my life working to help protect the world's environment.

'The possibility of a major warming in our lifetimes has to be taken seriously,' wrote John Gribbin, the editor and chief contributor to *The Breathing Planet*. And while the models they used back then were crude, and much uncertainty remained, Gribbin maintained that 'we should not commit ourselves to a course of action without being sure it is safe; the legal maxim "innocent until proven guilty" has to be reversed when considering the prospects of possibly permanent damage to the environment.'[10] He was invoking the precautionary principle, the same concept enshrined in the *Great Barrier Reef Marine Park Act* in 1975, the same principle that forty years later the dissenting senators would recommend be rolled back in their addendum to the report on the farm practices and water quality inquiry.

The year 1990 was the warmest year yet on record, as was 1995. In 1997, again the hottest year yet, the world finally understood that it needed to act. A total of 192 countries, including the United States and Australia, signed the Kyoto Protocol, committing themselves to limiting and reducing greenhouse gas emissions in accordance with agreed individual targets. The protocol set binding emissions reduction targets for thirty-seven countries, and targeted a modest 5 per cent emissions reduction compared to 1990 levels over the five-year period 2008 to 2012.

I can still remember the day I heard the news – that feeling of relief, of confidence in our international system, a belief that human beings could do anything if we rose above our differences and committed to action for the common good. It was a good day.

And then in 1998, a strong El Niño year, global average temperatures jumped to an all-time high, eclipsing the previous record by almost a quarter of a degree Celsius.[11] Reefs around the world started bleaching and then dying.[12] Between mid-1997 and late 1998, reports of severe bleaching came in from most tropical regions, including the Caribbean, the Atlantic Ocean, the Red Sea, the Arabian Gulf, the Indian Ocean, South-East Asia and across the Pacific. In February of 1998, extensive bleaching and some coral mortality was reported across the central GBR.[13] Bleaching on the Great Barrier Reef had been reported before 1998, but only in small isolated pockets and only on a small number of reefs.[14] This was something new. According to the AIMS long term monitoring team, 'Bleaching appeared to be most severe on inner-shelf reefs, least severe on outer-shelf reefs off Townsville and Cairns, and most pronounced in shallow parts of affected reefs.'[15]

The following year, Greenpeace commissioned Hoegh-Guldberg, who by then had become director of the Coral Reef Research Institute at the University of Sydney, to report what was now widely seen as the first global mass bleaching event – the bleaching of a large proportion of corals over large geographic areas. His findings were stark. Because of the underlying warming of the planet, and the continued accumulation of greenhouse gases in the atmosphere, bleaching would very likely become an annual occurrence on most tropical reefs around the world within the next thirty to fifty years. His warning was direct and bleak. If nothing was done to change the trajectory of warming, it would mean 'catastrophe for tropical marine systems everywhere', resulting in 'the complete loss of coral reefs on a global scale.'[16]

In an interview with journalist Marian Wilkinson, Hoegh-Guldberg shared his feelings. 'I remember exploring the risks of bleaching and mortality using a big Excel file and the answer emerging, which wasn't the answer anyone wanted. It wasn't going to be hundreds of years into the future. Rather, annual bleaching, back-to-back bleaching would occur around mid-century. I thought I must have made a mistake. I remember ... pulling my chair back from my desk and saying "Oh my goodness".'[17]

The reaction to his report was immediate and strong. Many scientists simply didn't believe his results. How could something as huge as the Great Barrier Reef just disappear? He was called an alarmist, and was attacked by climate sceptics, right-wing pundits and the Murdoch press. But then in 2002, the second-warmest year on record to that point, it happened again. The southern section of the GBR bleached, although not badly enough to cause extensive coral loss. Still, it was now becoming clear to anyone paying attention that this was not an isolated phenomenon, some freak occurrence. Hoegh-Guldberg had shown us a picture of what the future might hold, and it wasn't nice to look at.

~

When the Great Barrier Reef was established as a World Heritage site in 1981, many in Australia were justifiably proud. Recognising the long-term implications, the GBRMPA started working in the early 1990s on an ambitious twenty-five-year strategic plan for the park, the central goal of which was to meet the requirements of the World Heritage Convention.

In 1992, the High Court ruled in favour of Eddie Mabo and four other Meriam people from the Murray Island group in the Torres Strait, who ten years earlier had claimed traditional ownership of their islands. It was a landmark decision in Australian history. The High Court determined that the concept of *terra nullius* invoked by British settlers – the idea that the continent was essentially vacant before the arrival of Europeans – did not apply. That same year the Commonwealth passed the *Native Title Act*, enshrining native title rights in law. Henceforth, original inhabitants of Australia, people like Bob Muir and the remnants of his Woopaburra people, could claim native title of their historic lands.[18]

The GBRMPA was quick to realise the significance of the Mabo decision. Its strategic plan for 1994–2019 set out a vision for a World Heritage Area characterised by a healthy environment, integrated management, sustainable multiple use, maintenance of values, science-based decision-making using the precautionary principle, and involvement of the community – including Indigenous peoples.[19] The accompanying corporate plan for 1994–99 stated explicitly that GBRMPA would 'provide recognition of Aboriginal and Torres Strait Islanders' traditional affiliations and rights in management of the Marine Park'. For the first time, Indigenous people were invited to share in the management of their sea country and its resources. It was a big step forward. 'The Mabo decision was a major turning point for us,' Bob Muir told me. 'It's led to a lot of change. It was what motivated me to stick that flag in the sand on my country. It wasn't just for his mob; it was for all of us.'

As pressure on the reef built, the new park management authority increasingly began referring to the reef's status as a United Nations heritage site. Management efforts were cast in UNESCO's language of ensuring that World Heritage values were maintained, and that the Great Barrier Reef would be protected and preserved for future generations. Public education programs were launched, largely aimed at recreational users of the World Heritage Area, pointing out the damage done by coral collecting, the trampling of corals, anchors, overfishing and waste discharge. New efforts were put in place to protect endangered species, including dugongs, and green and hawksbill turtles, whose numbers had continued to decline.[20]

And then in 1997, a new report was released that explicitly assessed how Australia was doing in protecting the GBR World Heritage Area's outstanding universal value. Commissioned by the reef authority, the Queensland Government, and the Commonwealth Government's World Heritage Unit, and led by Percy Lucas, the report gave Australia a passing grade, with a warning. Lucas and his team from James Cook University concluded that while World Heritage status continued to be justified, more work needed to be done to ensure that the reef was protected.[21]

That same year, another Senate inquiry – this time on marine pollution in Australian waters – was completed. Its report singled out the Great Barrier reef for special attention. It acknowledged the reef's superlative beauty and diversity, but declared that it was now clearly 'at risk'. It found that just since the 1950s, when Queensland introduced its 'develop the north' campaign, flows of sediment and nutrients to the reef lagoon had increased significantly, largely from grazing and cane growing. Marine pollution was increasing rapidly, increasingly carrying harmful chemicals to the inshore reef lagoon.

By the early 2000s it was clear that despite the authority's efforts, the threats to the reef were continuing to grow. In 1999, Queensland cleared more land than all of the other states in Australia put together – over 400,000 hectares, the equivalent of almost 4000 Australian Rules football ovals each week. Since the start of widespread colonial development in 1860, more than half of Queensland had been stripped of its original native forest cover for agriculture.[22]

The rest of the world, too, was now recognising similar problems. The Worldwatch Institute catalogued similar crises of eutrophication from excessive runoff of sediment and nutrients in China, Japan, and the Black and Baltic seas. Runoff from the Mississippi Delta had created a biological 'dead zone' in the Gulf of Mexico the size of New Jersey, wiping out bottom-dwelling marine organisms, forcing the fishing industry to move further offshore in search of a catch.[23]

In a way, nothing had changed from the early decades of the 20th century: everyone wanted a piece of the reef and its adjoining coast, and the pressures just kept on mounting. And now, the dark thunderclouds of climate change had appeared on the horizon. The World Heritage status that had once been a source of pride for Australia and its government was about to become a nasty headache.

~

The UN's Intergovernmental Panel on Climate Change, the IPCC, issued its third assessment report in 2001. I still have my copy. The report provided a synthesis of the state of knowledge about climate change, based on a comprehensive assessment of peer-reviewed science by scientists from around the world.

According to the report, the last decade of the 20th century had been the warmest on record. Global average surface temperatures had increased by about 0.6 degrees Celsius since 1900. Humans were pumping more greenhouse gases into the atmosphere than ever before. New and stronger evidence had come to light that most of the warming observed over the last fifty years was due to human activities. All projections pointed to increasing temperature and rising sea levels over the coming decades, and the effects of global warming would persist for centuries.[24]

The impacts of these changes were everywhere to be seen. Snow cover and ice extent had decreased since the mid-20th century. The ranges of plants and animals were shifting poleward. Temperature-driven changes were being observed in all ecosystems, including marine environments. Rainfall and runoff patterns were changing. And according to the projections, more was to come. In the coming decades, if emissions were not significantly reduced, droughts and floods would continue to intensify. Maximum temperature records would continue to be broken. Wildfires would become more frequent and more intense. And more people would be harmed.[25] For anyone following even the most basic science, by the third such report, the language was becoming frighteningly consistent, and the predictions increasingly certain.

In January of 2001, George W Bush was sworn in as president of the United States, and he and his vice-president, the oil executive Dick Cheney, set about pulling the United States out of the Kyoto Protocol. By March, it was done, leaving the agreement in tatters, and any real hope of significant emissions reduction adrift. Then that November, John Howard was re-elected for a third term as prime minister of Australia. Many in his conservative Coalition government were anxious about what the Kyoto agreement might mean for business and the fossil fuel industry. Emboldened by success in the United States, climate sceptics in Australia started pushing for a similar result at home. Australia's so-called climate wars were well and truly underway.

In the wake of the 2002 bleaching on the GBR, the reef authority was already hard at work on an ambitious new zoning plan for the reef. By then, many in the broader community had started to recognise the emerging threat that climate change posed to the reef. The GBRMPA could not directly affect global greenhouse gas emissions, but it could drive improvements that would bolster the reef's overall health and resilience, making it better able to survive in a warmer world. The idea was to recognise the multiple uses of the park, balancing economic interests with conservation. The latest science was used to identify areas of outstanding value so they could be provided enhanced protection, while keeping vast areas of the park open for commercial fishing, recreation and tourism. Seven zones were included in the plan, which was approved by the Howard government in 2003. Highly protected zones were identified for conservation and habitat protection, where commercial fishing was banned and other activities allowed only with permission from GBRMPA. No-take zones were set aside.[26]

Adam Smith is a marine scientist with over thirty years of experience, and has been passionate about the oceans since he was a small boy growing up in New South Wales. Tall, genial, with a welcoming and slightly destabilising gaze – one eye is reef lagoon blue, the other coral brown – Smith was at GBRMPA when the zoning plan was being worked out. I spoke to him about the experience. 'There was lots of public consultation on the plan,' he told me. 'And there was lots of tension. Scientists, fishers, tourism operators, government, everyone had an agenda. But in the end, we got a good outcome, I think. No one was ecstatic, but everyone could live with it.' According to Smith and many others I spoke to, GBRMPA's zoning plan was a great success. 'This was transformational change,' said Smith, 'not incremental. Overall protection on the reef went from about 4 per cent to over 34 per cent. It literally set the global benchmark.'

The zoning plan helped to reduce the pressure on the reef from things like tourism and fishing, and in this sense, it was a great success, and a big step forward. But the plan did little to protect the reef from land-based runoff, and nothing directly to protect it from the emerging dangers of planetary warming.

In 2007 the IPCC issued its fourth assessment report, featuring Ove Hoegh-Guldberg as one of its authors. The news was not good. All of the findings and warnings of the 2001 IPCC report were reinforced. Emissions had continued to rise, and with them, temperatures and sea levels. Importantly, the report linked warmer ocean temperatures to mass bleaching and coral mortality, reiterating Hoegh-Guldberg's prediction of back-to-back bleaching events worldwide by mid-century. Not only had the world failed to act in the intervening six years, we were accelerating in the wrong direction.[27]

In that same year, Sir Nicholas Stern, a former chief economist of the World Bank and head of the United Kingdom's Government Economic Service, published his seminal work *The Economics of Climate Change: The Stern Review*. His findings were sobering. If the world failed to act soon to curb emissions, we faced the prospect of a permanent reduction of GDP of between 5 per cent and 20 per cent. Such a decline would make the Great Depression look like nothing more than a temporary inconvenience. The implications for the human race could be catastrophic. In contrast, the cost of reducing greenhouse gas emissions to a level that would avoid the worst impacts would be about 1 per cent of global GDP each year.[28]

But as momentum for action on climate change seemed to be building around the world, so too was the opposition. In Australia, climate sceptics like Bob Carter and his understudy Peter Ridd were embraced by the Institute for Public Affairs and its offshoot, the Australian Environment Foundation. Carter was asked to co-author a rebuttal to the Stern Review, and gave evidence at the US Senate Committee on Environment and Public Works at the invitation of Republican senator James Inhofe, a vociferous climate denier. Inhofe had famously described climate change as 'the greatest hoax ever perpetrated on the American people'.[29] At the hearing, Carter let fly with an all-guns attack on the IPCC report. According to Carter, the whole idea underlying the report – that global greenhouse gas emissions were linked to planetary warming – was just plain wrong.[30]

I remember reading the Stern Review cover to cover, and thinking, *Wow, if this doesn't convince everyone, I don't know what will.* If for no other reason than rank self-interest, we needed to get emissions down. We would actually all be significantly better off for it. At the time I had just moved to Australia with my family to take up a new role as head of sustainability for WorleyParsons, the big Aussie engineering company. Working under Peter Meurs, the visionary and influential co-founder of the company, we began developing a new way to embed sustainability, energy efficiency and emissions reduction into major project design and delivery from the outset, rather than as an afterthought or a bolt-on, as was so often done. In 2006 I had taken colleagues from WorleyParsons to see Al Gore when he came to Perth with his *An Inconvenient Truth* talk. You could sense the momentum. Industry was listening, and so was government. Finally, things were about to change.

One of my first projects at Worley was helping Woodside Petroleum design a new LNG plant to be located on Scott Reef, about 300 kilometres off the Western Australian coast in the Indian Ocean. Far from the coast, protected from the depredations of tourism and poor water quality, Scott Reef has long been considered one of the most pristine and spectacular coral reefs in the world. But like other reefs around the world, Scott Reef had been badly hit by the 1998 mass bleaching event, losing over 90 per cent of its coral. By the time I arrived in Australia, it had started to recover strongly. AIMS had been hired to survey the reef and, using their data, we showed the development partners that the risks to Scott Reef itself and the resulting economic costs and reputational damage of a spill or other mishap were so significant that a land-based solution was preferable. It was a great win for Woodside and for Scott Reef, but as we'd seen, protecting reefs from development pressure wasn't much help when a serious marine heatwave hit.

But by mid-2007 we were careening towards another crisis that would change everything, again of our own making. The highly leveraged mortgage market in the United States was unravelling, and the first signs of panic among financial institutions were being seen. I remember meeting in London with my brother, then a banker with one of the big investment houses, when the Bank of England agreed to bail out mid-sized UK bank Northern Rock. 'Everything is freezing up,' he told me. 'It's like being at a party. Everyone is just standing there with their glasses of expensive French champagne looking at everyone else, making sure that they are not the last one to leave. I don't know when the crash will come, but it's going to happen. Everyone knows it. If you've got any investments, get out. Get into cash. We all are.'

In the end, the Great Financial Crisis of 2007–08 cost the world's economy about 5 per cent of GDP for eighteen months. Thousands of unfortunate people lost homes, jobs and income in the crash. By 2009, however, global GDP had fully recovered and was already US$2.5 trillion greater than before the collapse, thanks to the strong intervention of central banks.[31] And yet, compared to Stern's warnings about the permanent costs of climate change, it was, as my dad used to say, 'peanuts'.

For the fight against climate change, the financial crisis was a huge setback. Getting the economy back on track became the focus for governments worldwide, and once again, meaningful action on climate change would have to wait. Opponents of climate action all over the world took advantage of the financial crisis to redouble their attacks.

In Australia, one of the loudest voices on the climate issue was columnist Andrew Bolt. In a full-page article syndicated around the country headlined 'Global warming – just settle down and wait for proof. After all, panicking is so undignified',[32] Bolt pushed one of the oldest climate denialist messages, *Don't act without incontrovertible proof* – the exact opposite of the precautionary principle enshrined in the *Great Barrier Reef Marine Park Act*.

In the article, he set out ten so-called myths, describing how these 'fears are being contradicted by the facts, and more so by the week'. Bolt's myths included that the world is warming. 'Wrong,' he claimed categorically. It had warmed between 1975 and 1998, he allowed, but since then temperature had dropped. Myth: The seas are getting hotter. Wrong again – if anything they were getting colder, he wrote. Myth: The seas are rising. Again, wrong. According to Bolt, the seas had been rising steadily for 10,000 years but had now stopped. Myth: The Great Barrier Reef is dying. Wrong. Bolt singled out Professor Hoegh-Guldberg, describing him as 'our leading reef alarmist and administrator of over $30 million in warming grants'.

Misunderstanding Hoegh-Guldberg's warnings, Bolt cited recovery from recent bleaching as proof that warming was not occurring, rather than as proof that the reef was still resilient, and could bounce back. He capped off the article by calling on readers to wait for proof, as if the mountains of evidence from scientists all over the world meant nothing, or did not actually exist. I still have a copy of that article, torn from a newspaper. The irony of Bolt's piece, read today, is haunting.

As if to cap off a decade that had started with such hope only to end in disaster, the GBRMPA released its first outlook report in 2009. In the authority's own words, 'The overall outlook for the Great Barrier Reef is poor and catastrophic damage to the ecosystem may not be averted. Ultimately, if changes in the world's climate become too severe, no management actions will be able to climate-proof the Great Barrier Reef ecosystem.'[33]

~

In 2010, the reef's uneasy relationship with the petroleum industry came full circle. Forty years after the battle over oil exploration under the reef ended, paving the way for establishment of the marine park, the Commonwealth Government approved an LNG processing plant on Curtis Island within the Great Barrier Reef World Heritage Area.

Concerned, the World Heritage Centre and the International Union for the Conservation of Nature prepared a State of Conservation report, which was presented to the World Heritage Committee in 2011. The findings were conclusive and damning for Australia. Oil and gas exploration and exploitation activities were 'incompatible with World

Heritage status'. Approval of the plant represented 'a clear potential danger to the property's outstanding universal value'.[34]

The committee expressed 'extreme concern' about the situation. A monitoring mission was recommended to determine whether the reef should be inscribed on the List of World Heritage in Danger. Australia was urged to undertake a comprehensive strategic assessment of the whole property and to develop a long-term plan to protect the property's outstanding universal value. Hereafter, the World Heritage Committee would be keeping a close watch over Australia's management of the reef.

The resulting report, prepared by the IUCN and delivered to the World Heritage Committee in 2012, called on Australia to ban all new port development outside existing major ports, conduct an independent environmental review of Gladstone Harbour and the Curtis Island development, and to substantially increase efforts to improve reef water quality. The central goal should be to protect the outstanding universal value of the reef. Australia was given until 2015 to present a plan to the UN, and was required to provide regular updates in the interim.

And then on 7 September 2013, Tony Abbott's Liberal–National Coalition won the federal election, ousting Kevin Rudd's fractious Labor government. I was at CSIRO at the time, running the national Climate Adaptation Flagship. I had known that Abbott had a strong anti-climate action view. In 2009 he had called the 'so-called settled science of climate change ... absolute crap'.[35] Killing off the Rudd government's carbon tax had been one of his major campaign platforms, along with stopping the flow of illegal immigrants into the country. Within weeks of the election, the words 'climate change' began disappearing from government websites and documents. Contracts for work on climate change-related projects were pulled or quietly wound up. The Climate Adaptation Flagship, intended to help the nation prepare for and cope with the changes that were looking increasingly certain, ceased to exist shortly thereafter.

Not surprisingly given the turmoil, Australia's first progress report back to the World Heritage Committee in 2013 was not received well. It found that only limited progress had been made, and remained concerned about ongoing coastal development. If Australia did not or could not improve the situation, the committee would once again have to consider placing the reef on the 'in danger' list. The consequences of such a listing for Australia could include global humiliation, a decline in the number of tourists coming to visit the reef, with knock-on effects on the economy of northern Queensland in particular, and increased political pressure from citizens who expect their government to protect national treasures.

But progress was being made. That year saw the publication of the Scientific Consensus Statement on reef water quality, which noted declining trends in reef ecosystem condition due to continuing poor water quality, and notably, 'cumulative impacts of climate change and increasing intensity of extreme events'.[36] In addition, a new Reef Water Quality Protection Plan was endorsed by the Australian and Queensland governments, with a goal of ensuring that by 2020, the quality of water entering the reef as runoff should have no negative impact on the health and resilience of the reef. The plan called for a 50 per cent reduction in anthropogenic dissolved inorganic nitrogen loads to the reef, a 20 per cent drop in sediment loads and nutrients, and an ambitious 60 per cent reduction in pesticide loads.[37]

Throughout 2014, the Queensland and Commonwealth governments were hard at work hammering out a plan that would safeguard the reef and meet the requirements of the committee. In March of 2015, the *Reef 2050 Long-Term Sustainability Plan* was submitted. It was a major step ahead. The plan galvanised governments around 151 actions designed to preserve the reef's outstanding universal value – everything from water quality improvement to better park management. And while no effort was made in the report to prioritise the actions according to the levels of risk or impact, the threat of climate change was not addressed at all.

Nevertheless, the Reef 2050 Plan was a success. It was welcomed by the committee, who noted with approval major new pollution reduction targets to improve water quality, restriction of port development and capital dredging to four existing major ports along the coast, and a range of other new initiatives. Significant new investment was also applauded, including $200 million to accelerate actions to meet water quality improvement targets. But there were provisos. The committee noted that many of the actions were the responsibility of the Queensland Government and were yet to be implemented. Australia was to report back to the World Heritage Committee in 2016, and then again before 2019.[38]

A journey that had begun with concern over the Curtis Island LNG development – which went ahead – had ended up with a major new conservation plan for the reef. As with Scott Reef in the west, petroleum industry activities – dredging, pollution, port construction and associated ship traffic – were seen as incompatible with the long-term health of the reef. But it was now becoming clear that the industry's product – the fossil fuels which were driving climate change – was fast becoming an even bigger danger.

In December of 2015, the world tried again to agree on a way of reducing emissions. The Paris Agreement was adopted by 196 countries at the UN Climate Change Conference COP21. The stated goal of the agreement was to hold the increase in global average temperature to well below 2 degrees Celsius above pre-industrial levels. Australia signed on, pledging to reduce its emissions by 26 to 28 per cent on 2005 levels by 2030, a commitment seen by many as far from adequate.[39]

Few on the World Heritage Committee probably realised it at the time, but El Niño conditions were already brewing in the Pacific. The previous two strongest El Niños of the 20th century were in 1982–83 and 1997–98 – the latter of which had caused so much concern on the GBR. At the time, each was considered a 'once-in-a-century' event.[40] The El Niño taking shape in 2015 was already threatening to be in the same class as those of 1982–83 and 1997–98, and would help drive the 2016 bleaching event on the Great Barrier Reef. Once-in-a-century events were now becoming once-in-a decade events.

14

Situation Critical

In 2020, in the wake of the just-concluded Senate water quality hearings, I was thrown into the middle of one of the most bizarre series of events I had ever experienced. The Senate report had not quelled the anger against the Queensland Government's new reef regulations. If anything, it had only exacerbated the situation.

The attacks on science continued. In another op-ed in *The Australian*, Peter Ridd reframed our Senate inquiry testimony – statements of fact already published in the literature and available to all – as 'remarkable confessions'. He even went so far as to call AIMS 'negligent'.[1] As a professional engineer, to see that word in print associated with my organisation was highly insulting. And so that day in 2020 when I received an email inviting me to a so-called 'duel' to settle the science of the reef, I was not surprised.

What did surprise me, though, was the program for the duel, which was already up on Facebook. The invitation included date and time, venue and speakers. A moderator had been invited and the format of the debate was set. The eminent Professor Peter Ridd would present one side of the argument, and I, the CEO of AIMS, would present the other.

The whole thing had been set up without my consent or permission.

My photograph was already up on the program. I remember standing there in my office, not quite believing what I was seeing, thinking, *Who does this?* To say that I was pissed off is to put it politely.

In my written response to the organisers, I expressed surprise at receiving 'an invitation to an event at which I was to be the main attraction', and asked that advertising for this 'non-event' cease, and that my photo be removed from its Facebook page. A duel was not the way to conduct meaningful scientific discourse. As an alternative, I suggested that he share his original research, data and analyses backing up his claims about the reef at one of the many scientific conferences, symposia, or in any of the peer-reviewed journals that are available to all scientists. These, I suggested, would bring his claims to a much wider audience and subject them to the same scrutiny that AIMS's science was subject to every day.

One of the world's most prestigious coral reef conferences, convened every four years by the International Coral Reef Society (ICRS), was at that time accepting abstracts for consideration. I suggested he participate. The conference would be the perfect place to present his analysis, and have his work scrutinised and debated by experts from around the world, in a constructive and collegiate forum focused on science, not opinion. At that point, AIMS scientists had already submitted over thirty abstracts to the conference.

But it was a forlorn hope. The pressure continued, including radio advertisements run in North Queensland spruiking the duel and my involvement. Despite the fact that I had declined to participate, the so-called duel went ahead anyway, livestreamed on Facebook, with Ridd debating a robotic effigy of me in front of a sparse crowd of friends and supporters in what I can only describe as a one-sided circus stunt.

It didn't end there. Not too long after, after a particularly intense week with RRAP taking up the majority of my recorded sixty-two hours of work, Heidi and I decided on a morning paddle on the Ross River and lunch with a couple of margaritas – iced, traditional, no salt – at a Mexican restaurant in town. After a great day, feeling relaxed and happy, we were returning home, approaching the main gate to the AIMS site on Cape Ferguson. Normally open during weekday working hours, the gates are closed on the weekends. A couple of vehicles were pulled over at the side of the road just outside the gate, and some people were milling about.

As we got closer, we could see a man with what looked like a movie camera filming the front gate. A woman stood at the side of the road peering skyward, working some kind of gaming console with her hands. I slowed the car and stopped beside a diminutive older man standing on the shoulder, facing away from the road. I rolled down the window.

'Excuse me,' I said. 'Can I help you?'

The man turned around and we stared at each other. It was Ridd.

'What are you doing here?' I asked.

Ridd stammered a moment, and said something like: 'You wouldn't let us in, so we're filming here.'

I could feel my anger rising. All the weeks of personal attacks, and him here outside my home on a Sunday afternoon. 'I tell you what,' I said, fixing him with a stare. 'You wanted a duel. How about we stand over there on the other side of the road and have it out right now, just you and me?'

Ridd blanched and said something like: 'Don't be ridiculous.'

Before things could escalate further, Heidi put her hand on my arm and gave me that look that meant I needed to cool down. Rather than pushing it any further I politely asked him to get out of my face, rolled up the passenger side window, swiped my card, and drove away, making sure that none of his party came through the gate after us.

That things had got quite this comical, quite this ridiculous, was sad testament to how fractious the reef discourse had become.

~

Through the rest of 2020 and into 2021, scientists kept working to build our knowledge about coral reef ecosystems, the threats they faced, and increasingly, how to counter those threats.

Long-term monitoring of the reef in 2021, another La Niña cooling year, showed strong recovery of coral cover in all three regions during a twelve-month period of low disturbance. The reef was bouncing back. It was great news. But the data also showed the first hints that species composition might be changing as a result of the accelerated cycles of loss and repair. Recovery seemed to be dominated by fast-growing species like the branched *Acropora*, which are much more vulnerable to bleaching and physical damage from storms and cyclones.[2]

Predictably, climate deniers jumped on the new monitoring data as proof of what they had been saying all along. The reef was just fine. AIMS science was suddenly trustworthy enough to quote with confidence to support this view.

But while the short-term news was good, the long-term prognosis for the reef remained poor. A new study by a team of scientists from around the world, including the United States, France, the United Kingdom and Australia, found that coral calcification rates globally would decline severely over the coming decades under global warming and ocean acidification. Published in one of the world's most prestigious scientific journals, the work confirmed the warnings provided by the De'ath 2005 paper so often questioned by reef contrarians. The team cited coral bleaching as the prime cause of the decline. By 2050, under business-as-usual emissions, most reefs around the world would be eroding rather than accreting, slowly sinking beneath the rising seas.[3]

Spurred on by an overwhelming sense of urgency, the multi-institution RRAP team continued its work. Money was now flowing from the Reef Trust Partnership through the Great Barrier Reef Foundation, and the first scientific breakthroughs were starting to occur. We likened the effort to the first moonshot: complex, stretching the limits of science and technology, and inherently uncertain.[4] The US National Academy of Sciences convened a panel of international experts, including RRAP team members, to develop a blueprint for coral reef survival in the face of climate change. They found that existing conservation measures such as zoning, marine protected areas and fisheries management – the kind of things that were already being done extensively on the GBR – were no longer sufficient to protect reefs. New methods to build reef resilience were needed.[5]

Helping reefs adapt to warming oceans was increasingly being seen as a complex challenge, involving potentially dozens of different technologies working in combination over varying times and scales. The team was starting to realise that when the decision to intervene was finally made – probably in the face of extreme reef loss after a series of back-to-back bleaching events leaving the reef no time to recover – we were not going to be able to save everything. There would be neither the money nor the time, nor the available resources. Hard choices would have to be made, and we needed a decision-support process to inform those decisions.[6]

For many scientists, this kind of thinking not only illustrated how desperate things had become, despite the benign conditions on the GBR over the last year, but also showed the widening gap between what we knew was increasingly likely – the eventual loss of much of the reef – and the message being hammered home by the opponents of action to address climate change. Many of my colleagues found the situation profoundly confronting. That we were even researching this question of which parts of the reef to save and which to let die – that it had actually come to this – was bad enough. But to be continually pilloried in the press by people who had very little understanding of what was actually going on, and to think that some part of the public was listening to them, was deeply troubling. We knew we had to do more of everything and do it faster.

The pressure had mounted further in November of 2020, when the IUCN released its third global World Heritage Outlook Report. In a shock to many in government, the GBR was further downgraded from its 2017 status of 'Significant Concern', to the 'Critical' category. The decision was officially attributed to the lack of progress towards Reef 2050 goals, and further decline to the reef's outstanding value. Iconic species such as loggerhead, hawksbill and northern green turtles were in decline.[7]

And so, in 2021 when the IUCN recommended to the UN World Heritage Committee that the GBR be put on the 'in danger' list, it was not much of a surprise. The Australian Government's continued lack of progress on meaningful emissions reduction was clearly a factor. At that stage many stakeholders were of the opinion that the Reef 2050 Plan vision was no longer attainable because of the now-apparent dangers of warming, and the decision had been made to redraft it. Climate change and Australia's commitments under the Paris Agreement to were to be mentioned explicitly.

In my view, which I shared with anyone who asked, Australia was being punished for its lack of progress on meaningful emission reduction, and the reef's heritage status was being used as the stick. Monitoring data over the last couple of years had shown good recovery,[8] an indication that despite the clear need for improvement in some key areas, the GBR

World Heritage Area was indeed being well-managed compared to most other reefs in the world. If climate change really was the single biggest threat to the reef – and it clearly was – then all twenty-seven reefs on the UN's World Heritage List should be placed on the 'in danger' list. After all, as the UN itself maintained, the 'in danger' listing was not meant as a punishment, but to bring attention to the need for determined action. Climate change was a global problem, and as Ove Hoegh-Guldberg's report to the World Heritage Committee had shown, all the World Heritage reefs were facing the same problem. Australia was being singled out because of our emissions record. In my opinion the UN was using the wrong hammer for the right nail.

Some agreed with that view, including the government and many in the tourism industry, who were worried that an 'in danger' listing would signal to the world that there was no longer anything worth seeing.[9] Paradoxically, I found myself on the same side of this particular argument as many of the climate deniers, but for different reasons. To them, the 'in danger' listing was not justified because the reef was fine, and climate change was not an issue.[10] To me, climate change was exactly the issue, and as a global problem, needed global solutions. Put all the heritage reefs on the 'in danger' list – now that would bring some attention to the issue.

Many did not agree with me, including some opposition MPs, well-known scientists like Professor Terry Hughes, and some conservation groups. To them, the ends justified the means. If the 'in danger' listing could be used to force the government into stronger action on climate change, and on reef protection in general, then it should be invoked.

One of those non-governmental organisations that supported the listing was the World Wide Fund for Nature. I met with its long-serving Australian CEO, Dermot O'Gorman, in a cafe in Sydney in late 2023. 'The Great Barrier Reef is a global icon,' he said. 'It has a special place in Australian hearts, and the world's. Because of this, it has been on the frontline of innovative legislation and policy development for a long time. Water quality improvement is a work in progress, but over the years we have managed to stop oil and gas development, ban dredging and dumping, and now we will shortly end gillnet fishing on the reef. We've done a lot that science says we need to do to build the reef's resilience.' Our conversation turned to climate change and its effects on the reef. 'The reef is a place where we can fight politically and economically,' he said. 'It has leverage because people value it. Australia and Queensland need to get out of fossil fuels by becoming renewable energy superpowers.'

In the end, after some intensive lobbying from the Australian Government, and despite the support of WWF and others, the twenty-one-country World Heritage Committee rejected UNESCO's recommendation.[11] Instead, UNESCO was asked to carry out another reactive monitoring mission in early 2022. Environment groups were furious, claiming that the decision was the result of cynical lobbying.[12] The reef was saved from the 'in danger' listing, and Australia was off the hook, again – for now.

15

Acts of Resistance

Through 2021 and 2022, Heidi and I continued our evening walks along the AIMS beach. We picked up plastic, and the plastic kept on coming. We noted the tracks of turtles coming ashore to nest, and the footprints of the animals who destroyed their nests. And we learned about the story of our site's most famous European castaway, whose initial shelter in the rocks at the southern point of our beach we passed by most days.

In 1846, in the season when the sun's sharpest edge had started to dull and the oppressive humidity of the previous months had finally begun to relent, the people who had lived for a thousand generations in the area around what was to become Townsville, the country where the three big rivers met the sea, observed streaks of fire falling from the sky towards the coast. The people believed that these falling stars were a sign of danger, and the comets were the ghosts of ancestors trying to find their way back to the world.[1]

The elders sent scouts to investigate this strange occurrence.

They found the first of the strangers stretched out on the sand near one of their own boats they called *woolgoora*, skin burned, lips and tongue swollen from thirst, his pale eyes staring up at the sky. It didn't take the scouts long to find the others. They found two more dead and then four more alive, huddled in a cave in the rocks by the sea. The strangers – two men, a boy and one woman – were in bad shape, pale and bone-thin, thirsty and hungry. The scouts provided the newcomers with food and water.

A messenger was sent back to inform the elders, and the next day a large group arrived from the camp to wonder at these new arrivals. The woman and the older man refused to shed the dirty rags they covered themselves with, but the boy and the younger man quickly learned the ways of the people.[2]

A year later the boy and the older man died, and the woman soon after. But the young man grew strong and as the days passed the people considered him as one of them, speaking their language fluently, hunting with the men, and even, after some time, taking a wife.

And then in 1863, seventeen years after the barque *Peruvian*, on which he was a crew member, ran aground near Minerva Shoal on its way from Sydney to China, James Morrill walked into a sheep station near the Burdekin River. His time with the people was over.

Back in 'civilisation', which by now was encroaching deeper and deeper into Aboriginal lands, Morrill watched in horror as the violence between the colonists and the so-called savages escalated into open warfare. He offered his services to help broker a ceasefire and end the war. But his ideas were met with disdain by many colonists, and his life was threatened. The colonists wanted the land, and they wanted the Aboriginal people gone.[3] James Morrill never did broker peace. He died in 1865, just two years after leaving the people who had rescued him, seen him back to health and welcomed him as one of their own.

To Heidi and me and many of my colleagues, Morrill's story seemed to encapsulate so much that we were seeing on the reef. Despite the progress, and the efforts of so many dedicated people, the reef was in greater peril than it had ever been. Morrill's story reinforced the dominance of the European-Western ethic, its system of economics, and its attitudes towards nature and towards the country's original inhabitants. Morrill's shelter was a daily reminder of the stark contrast between the pristine land and seascapes of that earlier time and their present state of degradation. And it reminded us of the seeming lack of fundamental gratitude for what we had inherited, and of our responsibility to the future for its protection.

It was this, in part, that catalysed our decision. By late 2021, the idea we'd had of doing something to protect the turtles that still came to nest on the beach near Morrill's cave was taking shape. With the help and guidance of the Bindal people, the traditional custodians of the lands of the AIMS site, a small band of AIMS volunteers identified and protected ten nests. In that first year of what we now called Operation Gungu – the Bindal word for turtle – we watched four nests hatch successfully. We were lucky enough to actually have had a team present to watch some of the small baby turtles flipper their way to the surf and swim away, and capture it on video.

At the end of the summer, we dug up the remaining protected nests to determine why they hadn't hatched. By carefully recovering the eggs and inspecting them, we could see that our protection devices had worked. They had prevented predators such as feral pigs, foxes and dingoes from eating the eggs. The embryos had started to develop. But then at a certain point, development had ceased, and the eggs had simply cooked in the sand. The temperatures inside the nests had been too high.

Despite this, we celebrated our limited success. For the first time in many years, baby turtles were leaving this beach to continue the cycle that had been going on for over a hundred million years. It was hopeful, uplifting stuff.

And then one day I was talking to one of our scientists, an expert in marine megafauna, including turtles. The annual congregation of endangered green turtles on Raine Island in the GBR was going on, and the news had featured stunning video of tens of thousands of turtles congregating in the sea around the island, waiting for the moment when they would return to the sands where they were born, to nest in their turn. It was a wonderful testament to the resilience of nature.

'So many of them,' I said to her. 'It's fantastic to see.'

'Yes, it is,' she said.

She did not seem very enthusiastic, and I asked her why.

'Well, the problem is, they are already extinct. They just don't know it yet.'

I was floored. The scientist explained to me that the vast majority of the turtles being born now were female, including most likely the turtles from our beach. Once the eggs are laid, it is temperature which determines the sex of the hatchlings. Temperatures over about 29 degrees Celsius produce females, while anything cooler than that produces males.[4] With climate change driving warmer temperatures, the feminisation of the turtle population was well underway and showed no signs of slowing down.[5]

It was hard to hear. Another example of the illusion of safety that is so easy to convince ourselves of, in the face of something as slow-moving and pervasive as climate change. We see lots of turtles, therefore they must be okay, doing well. We see coral recovering, so the reef must be fine, and the earlier losses must be just an aberration, or part of a natural cycle. There is nothing to worry about.

But it was precisely this simplistic thinking that was driving us towards catastrophe, step by step, sleepwalking towards the cliff. Make that a multitude of cliffs of different heights, a gradual unravelling, with just enough breaks along the way to convince ourselves that everything will be okay.

~

That southern hemisphere summer of 2021–22, the heat was back, despite a second successive La Niña. Temperatures soared, cooking the turtle eggs in the nests we had protected. We watched through Christmas and into the new year as sea surface temperatures rose and persisted, accumulating heat in the waters of the GBR. And then in early 2022 it happened again. Corals started bleaching all across the reef.

Aerial surveys were launched, and soon the message came back that another mass bleaching event was unfolding on the GBR. Bleaching was seen over a large spatial extent and varied in severity, with some reefs badly hit and others showing no signs of bleaching.[6] For the government, the timing could not have been worse. An election was due in just over a year, and with Australia's climate ambition still being questioned, the World Heritage Committee's reactive monitoring mission was to arrive shortly. Their task: to determine yet again if the GBR was to be listed 'in danger'.

As the bleaching was unfolding on the reef, Russia did what many would not have thought possible in the 21st century, and on 24 February invaded Ukraine. Heidi is Ukrainian Canadian; her family is part of the huge Ukrainian diaspora in Canada. In the weeks leading up to the invasion, Vladimir Putin had repeatedly denied he would invade, and had variously claimed that Ukraine was not a real country, that it did not deserve to exist and that Ukrainian was not a real language.

It hit me hard, all of it. Increasingly, I came to see the rise of autocrats like Putin, with their total disregard for the truth and singular pursuit of power and personal wealth, as somehow synonymous with and linked to climate change and the effect it was having on the natural world. I have always hated bullies and liars. To me, Putin and those who had been standing in the way of progress on climate change were part of the same problem. Protecting their wealth and status in the world was worth any price, including human lives or those of the other creatures who shared the planet with us. I had to do something.

Had I been younger, I would have gone to fight for Ukraine. I wanted to. But realism, guided by my wife, prevailed. Instead, I dusted off my high-school Russian, started learning Ukrainian (which is about 40 per cent similar), and applied to the Ukrainian Armed Forces for a press pass. If I couldn't fight, I would write – tell the stories that needed to be told. When my pass came through a few months later, I started planning. Over the last few years, I had accumulated a lot of leave. This, I was determined, was how I would use it.

And then, in May of that year, the landscape in Australia changed. In the federal election, the conservative Liberal–National coalition was swept from power, and a Labor government led by Anthony Albanese was elected. One factor in the election was the success of a number of so-called teal candidates – socially progressive conservatives disaffected by the Coalition's lack of action on climate change. The new government had campaigned on decisive climate action, including legislating national carbon reduction targets, and immediately set to work towards that goal. My new minister, Tanya Plibersek, was not only smart and dedicated to achieving real progress on the environment, she was a party heavyweight with real influence in cabinet. After a decade of political feet-dragging on climate, there was finally real hope that Australia would make some progress.

By this time, Lord Howe Island had reopened to visitors after the long period of Covid physical-distancing measures. Heidi and I spent a week there in October, hiking and cycling and swimming in the beautiful lagoon, home to the world's most southerly coral reef. Terns nested in their thousands, crowding the beaches. Windblown cirrus wisped across defiantly blue skies – the air there is as clean as anywhere on the planet. There on that tiny speck of rock in the middle of the South Pacific, the world's problems seemed very far away.

But reality returned soon enough, hard-edged and brutal. A few weeks later I took leave and left for Ukraine. Over the next month, I criss-crossed the country speaking to volunteers, refugees, internally displaced people, aid workers, doctors, soldiers and regular people. It was cold, and the Russians were bombing the country's power infrastructure in the hope that they could freeze the Ukrainians into submission. I spent a lot of time in bomb shelters and in candle-lit rooms and bunkers. The stories I heard were both frightening and uplifting. Some of the things I saw will stay with me for the rest of my life. I returned home a different person in a lot of ways, with an abiding respect for the Ukrainian people, and a deep sadness after having witnessed the waste and cruelty of war. There were so many problems the world needed to deal with. To see us wasting so much money and effort, and sacrificing so many people on a war of choice – one man's folly – was crushing.

Back home after a month that felt like a year, I found myself increasingly impatient with the lack of progress on the big issues facing the world, and disdainful of those who would not or could not see what was going on. The climate deniers and reef contrarians continued their attacks, kept accusing scientists and conservationists of conducting a 'scare campaign', howling about 'the gross misrepresentation of the state of the reef'.[7]

Did they honestly believe that thousands of scientists around the world were somehow involved in a giant conspiracy to hide the truth of the reef's robust health and vibrancy, to manufacture data to make it look as if the world was warming when in fact it was not? It seemed beyond comprehension. And yet the vitriol continued, fuelled by the politics of division and profit.

Despite this, the new government acted quickly on one of its major election promises and introduced a bill that would commit Australia to cutting greenhouse gas emissions by 43 per cent against a 2005 baseline by 2030, and to reaching net zero by 2050. It was a bold move whose time had finally come. In September, with the support of the Greens and the teals, the bill was passed into law. It was a big step forward for Australia.

In August of 2022, AIMS had released the annual long-term monitoring summary report on the health of the GBR. It showed that once again, the reef had dodged a missile.[8] Despite the soaring temperatures in another La Niña year, the weather had broken just in time, bringing cooler conditions, overcast skies and rain. Most of the coral that had bleached over the previous summer recovered. The central and northern regions of the reef recorded their highest levels of coral cover in thirty-six years. Only in the southern third of the reef did coral cover decline, largely due to outbreaks of crown-of-thorns starfish.

Despite the good news, the report again warned that much of the observed recovery had been driven by fragile fast-growing branching and table corals which were more susceptible to bleaching, easier for cyclones to damage, and happened to be the favourite food of the crown-of-thorns starfish. While the record coral cover levels was great news, and showed how resilient the ecosystem still was, it did not in any way mean that the reef was somehow okay. After four mass bleaching events in seven years, and the oceans sucking up record amounts of heat, the reef remained in peril. The message going forward was simple: we were now living in a world where the reef was at risk of severe bleaching in any given summer. How the ecosystem would respond, we did not know. We were entering the uncharted waters that Ove Hoegh-Guldberg and others had warned us about more than two decades before.

Of course, this was not how the contrarians and deniers saw it.

Again, when the science reported good news about the reef, it was to be congratulated, and was suddenly once again trustworthy. The record coral cover, to them, was proof of what they had been saying all along. Climate change was not happening. They saw the recovery as a long-term indication of the reef's robust health, rather than a testimony to its remarkable, yet finite, resilience.

The passing of the new climate bill was a major defeat for the climate wreckers. By then, it was abundantly clear that their one overarching goal was to prevent action, and if that was not possible, to delay it as long as possible, regarless of the consequences. Claiming that there was disagreement among scientists remained a key frontline tactic, one that had been scientifically demonstrated to create public confusion and disengagement.[9]

As I said to a colleague who describes himself as an 'informed climate sceptic' recently, there isn't a day goes by where I don't wish that the deniers were right. Life would be a whole lot easier. How wonderful it would be to wake up each morning without the heartsick worry about what I am leaving behind for my sons. But I have yet to see any convincing peer-reviewed science, or any science at all, that even begins to prove that climate change and the other major global changes are not happening, and that human activities are not responsible for the current rapid observed changes. Wishful thinking isn't going to solve the world's big problems. Sometimes, you have to fight.

~

And so, we kept battling.

All the climate modelling and the observed data was pointing to only one conclusion: the likelihood that the world was going to keep global warming below the 1.5 degrees Celsius target was now vanishingly small. And yet there was still a chance. And if it wasn't 1.5 degrees, then perhaps we could keep to below 1.75 degrees, or 1.9 degrees. We were entering the time when everything mattered. Every tonne of carbon emissions avoided, each ecosystem strengthened to the point where it was better able to survive, every tree planted, or better yet, left standing, was a win. And whatever we did, we were going to have to adapt.

The IPCC's sixth assessment report had been issued the year before, and for those following the science over the decades, the message was by now monotonously familiar. The accumulation of scientific work undertaken all around the world, by tens of thousands of scientists in dozens of countries, once again confirmed that the earth was warming, and it was clearly due to human activities. Each decade since 1850 had been successively warmer than the one before. The word the report used to describe these findings was 'unequivocal' – certain. For scientists, a notoriously cautious bunch, this kind of language implies confidence born from an overwhelming weight of evidence.[10]

Global consumption of fossil fuels had doubled in the last fifty years. The rate of deforestation, a major contributor to greenhouse gas emissions, had increased since 2015. As a result, the concentration of CO_2 in the atmosphere was higher than it had been for at least the last two million years, and methane concentration was the highest it had been for at least 800,000 years. These concentrations far exceeded any that could have occurred due to natural processes. It was clear that these changes were due to our activities.

And the results were there for everyone to see. Rainfall patterns were changing across the globe. Ice was melting, not only at the poles, but from the world's mountain glaciers. The upper ocean had warmed, and 2021 had seen a record increase in ocean heat content. As the oceans warm, they expand. Combined with the influx of water from land-based ice melt, it meant sea levels were rising. In fact, the rate of sea level rise had accelerated steadily since 1901, such that sea level was now one-fifth of a metre higher than it had been at the dawn of the 20th century. And all over the world, climate zones were shifting poleward, which meant that the ecosystems which had evolved in those zones had to follow, if they could.

The Great Barrier Reef is one of those ecosystems that cannot move poleward. The structures of the reef, in all their diversity, have been accreting and evolving for tens of thousands of years, in conditions ideally suited to tropical reef formation. The southerly limit of the GBR is determined by the availability of ocean-bed substrates suitable for reef growth. South of the GBR, there are very few places where the sea floor is shallow enough to accommodate reefs.[11] The GBR has evolved in exactly the right conditions – geological, biological, climatic and geographic – over tens of thousands of years. But the climate is now changing over a matter of decades, not millennia. The reef will not be able to move. There is nowhere for it to go.

Even the world's southernmost fringing coral reef at Lord Howe Island, over a thousand kilometres south of the GBR, has not escaped the effects of warming oceans. During the global mass bleaching event of 1998, bleaching was recorded at Lord Howe for the first time, but with little loss of coral. In 2010 the island's reef experienced its most extensive and severe bleaching event to date.[12] Coral communities there are at their extreme southern limits of distribution already, and include subtropical species which are non-existent on the GBR.[13]

Until recently, environmental conservation has been based on the principle of prevention: protect the ecosystem from impacts, limit access, restrict exploitation, and the resilience of the system will maintain overall function and productivity. But in the Anthropocene, this conservation paradigm has become increasingly untenable. Human impact on the earth's processes are now so deep that active efforts to help ecosystems adapt are no longer a matter of if, but how, and when.

It was now clear: the GBR would have to adapt if it was to survive the era of anthropogenic climate change.

And the Reef Restoration and Adaptation Program was making good progress towards its goal of ensuring that, if called for, science could deliver meaningful aid to the reef in time. The last two years of the program had been spent identifying prospective intervention techniques, and assessing their potential to be effective at a large scale. One of the approaches that had shown the most promise so far was to selectively breed heat-tolerant corals in captivity, and then place them on the reef in strategic locations where they could grow and spread their genes. Efforts to breed heat tolerance into several species of coral had been going on at AIMS since 2015, through the work of Professor Madeleine van Oppen. Van Oppen's work was focused on assisting the natural evolutionary processes she had observed occurring out on the reef. The emerging challenge was to determine whether these tolerances could be passed on to successive generations, and then to find out how to accelerate the process.

One of the leaders of this effort was Dr Line Bay. 'We've made huge progress,' Line told me in late 2023. 'You have to consider that when we started this, we realised it was ambitious, a global-scale effort. All we had were pieces of the puzzle, but now the bits are starting to fit together.' Soon, she said, her team would be ready to deploy the first batch of ten thousand protected, heat-tolerant corals onto a test reef. 'It will be a small step on the way to where we need to go,' she told me. 'But it's a start.'

Recent advances in our knowledge of coral reproductive biology were now allowing us to cultivate far more species than ever before.[14] We had successfully adapted several species to warmer temperatures and had begun testing their survival on the reef, paving the way for larger-scale field trials. The culturing process in the lab, however, was slow. As coral spawned in captivity – at exactly the same time as they did out on the reef – eggs and sperm were collected by hand by scientists armed with pipettes. It was a delicate, slow and tiring process, demanding a lot of patience and care. If we were going to culture millions of baby corals in captivity, we were going to have to find a better process. Part of the problem is that, in the hours after spawning, the fertilised eggs are extremely delicate, defying attempts to automate the process.

But work by the RRAP team had revealed that for the first forty-five minutes or so after fertilisation, eggs were actually extremely robust. If we could collect and mix spawn early, maybe we could speed up the culturing process. The team developed and successfully tested a new automated system for coral spawn collection, mixing and processing which increased the output and speed of coral aquaculture production by at least a hundred times.[15]

One of the other big challenges in coral reef regeneration had always been the low survival rate of baby coral. On the reef, as few as one in a million larvae might actually successfully settle, attach to the substrate and start to grow. Even then, the small nubbins are subject to predation, and many do not survive to reproduce. As Charlie Veron writes in his book *A Reef in Time*, coral larvae faced 'long odds' in their battle to survive.

To overcome this major problem, the RRAP team had been testing a variety of ways to get more of the heat-tolerant corals that had been bred in captivity to the places we needed them, faster and more easily, while significantly increasing survival rates. The most promising process involved a two-part settlement and seeding device system. Settlement was optimised on a carefully selected carbonate material favoured by the corals. The material was cast into sheets made up of dozens of easily-separated thumbnail-sized tabs, each now home to a comfortably settled baby coral. The whole process was simple enough to be automated, which would drive up production rates to the industrial scales needed. These tabs could then be loaded onto seeding devices designed to carry the baby corals down to selected places on the reef, and protect them against predation by nesting them within nooks and clefts on the device surface that prevent access by hungry beaks and mouths. Early designs had already shown a 50 per cent survival rate after six to nine months.[16]

Another key to adaptation success was the ability to predict where interventions should best be deployed, and how well they would function. Early modelling results had already shown that, even at a modest scale, well-targeted interventions could have significant impact over time. And the models were getting better all the time.

We realised that now, for the first time, there was real promise that we could, if needed, intervene on the reef to help it survive the coming storm. Hope is a powerful thing.

16

An Uncertain Future

At the start of 2023, I was nearing the end of my agreed term as CEO of AIMS. A lot had happened over the past five and a half years. I was now reporting to my eleventh minister, and with her help we were about to get the biggest funding boost in our history, ensuring that we could continue to deliver the reef science the country needed. Australia finally had real, binding climate legislation that committed us to doing our share to reduce emissions, in line with the goals of the Paris Agreement.

I'd seen three major bleaching events on the reef, and then watched the reef recover strongly under the cooling effect of an unprecedented run of three La Niña years in a row, and as many seasons without a major cyclone. The battles with the deniers and sceptics continued, but as the reality of what was going on became more and more evident, their voices were gaining less and less traction with the public.[1] And I had learned a lot about the struggle over the fate of the reef that had been going on for more than a century.

We'd done some great science, developed a strong culture of partnering with traditional owners, significantly improved our safety record and started construction on an expansion of the National Sea Simulator that would almost double its capacity. Along the way, we'd led the country's reef adaptation program to some major breakthroughs, despite the difficulties the Great Barrier Reef Foundation had experienced raising money in the first few years of the project. Covid had made fundraising significantly more difficult, especially from the large charities the foundation was depending on. So too had the fact that everyone now knew they had half a billion dollars in their bank account – 'Why do you need my contribution when you already have so much?' was the line. But perhaps most of all, the government's controversial decision to give that much money to the GBRF, without any sort of competitive process, had inadvertently damaged the reputation of the foundation, hindering its ability to fundraise. Only now, with painstaking effort, was the GBRF starting to recover, much like the reef whose name it included. The foundation was actively engaged in driving innovation and collaboration across the reef. Several big new donations had been identified, and new money was flowing through the foundation to organisations working on the GBR.[2]

Operation Gungu had gone ahead again that summer on AIMS's beach in Cape Cleveland, and we'd improved on the previous year's results. Twelve nests were protected from predation, and this time were shaded from the sun with simple sailcloth awnings. Ten nests produced hatchlings, doubling the previous year's success. Temperature loggers installed in the sand at the depth of the nests showed that the shading had worked, reducing average temperatures in the nests by about 1.5 degrees Celsius. By our best estimates, over 350 hatchlings made it to the sea, and based on the cooling we'd achieved, some of them would have been males.

No longer could we argue that our effect on nature was too small, the world simply too large, for we humans to have any effect. We were the ones who were putting these creatures at risk. We had to be the ones to help them. It could be done, and more importantly, it *had* to be done.

And then in March, the IPCC delivered the final part of its sixth assessment report. The UN secretary-general, António Guterres, called the report 'a clarion call to massively fast-track climate efforts by every country and every sector and on every timeframe. Our world needs climate action on all fronts: everything, everywhere, all at once.'[3] The report warned that the window of opportunity to stay below 1.5 degrees Celsius of warming was fast closing. Extreme weather driven by climate breakdown had already led to increased deaths from intensifying heatwaves in all regions, as well as millions of lives and homes destroyed in droughts and floods, millions of people facing hunger, and 'increasingly irreversible losses' in vital ecosystems.[4] Guterres's words sounded eerily similar to the 1987 predictions in *The Breathing Planet*.

Reading the report, the overwhelming sense I got was that after decades of warnings, finally, now was the moment. Right now. For years, too many of us had latched on to the deniers' illusions of false hope. Every time the trend wavered, even just for a few years, we'd heave a big sigh of relief, and say to ourselves, 'Maybe they're right, maybe it isn't happening,' and so absolve ourselves of responsibility and guilt. Even the tiniest sliver of doubt was enough. When the rise in global average temperatures slowed for a few years after the big 1998 El Niño, the deniers jumped. In an article for *The Telegraph* in 2006 headlined 'There IS a problem with global warming ... it stopped in 1998,' Bob Carter crowed triumphantly about the prediction of 'hypothetical human-caused climate change', as 'such deep dreaming, or ignorance of scientific facts and principles, that they are akin to nonsense'.[5]

Of course, we now know that Carter and his cohort were hopelessly and completely wrong. The slowing of warming between 1998 and 2012 was due to natural variability in the climate system, notably a change in ocean circulation patterns, which meant that more of the accumulated heat was being taken up by the oceans, leaving the atmosphere less affected.[6] After 2012, atmospheric warming accelerated again, and has kept on rising since, as greenhouse gas emissions have grown.

In a way, this was what was happening to the reef. Every time the natural system stabilised, or recovered after suffering damage, the deniers would cry foul. *Look*, they would say, *we told you so. It's all fine now, there's nothing to worry about.* But they failed to see, in the increasingly rapid phases of destruction and regrowth, the instability creeping into the reef system. Is it so hard to understand that you can lose half a living system in 2016–17 and that if conditions are good it can regrow, but still be in danger from the next big event? And if those events start coming more frequently, as they are, is it that hard to understand that the system might eventually reach a point where it would not be able to bounce back?

And yet at every step, the deniers latched on to any shred of evidence they could find to justify their position, oversimplifying it and taking it out of context, spreading confusion wherever they went. That they didn't even bother offering their own data or analysis was enough to reveal their real intent: to sow doubt in the minds of the public. That's all they had to do.

After a hundred years, we have learned so much about the Great Barrier Reef. We've mapped the reef in all its glory from north to south, unlocked the secrets of coral reproduction and growth, catalogued thousands of species of coral and fish and rays and all of the other creatures who live within the protective architecture of these undersea forests. We've seen first-hand the effects of warming on coral's unique symbiotic pairing, and explained the mechanism in great detail. We've created models of the reef's ecology which have allowed us to predict how the reef will change as our seas continue to warm. And all of this has been accomplished through the steadfast application of science, step by step over the years, each study building on the last, accumulating knowledge and understanding. It has been done through the testing of hypotheses by experimentation and observation, and by thoroughly checking the results through a comprehensive system of expert peer review.

When Maurice and Mattie Yonge arrived on Low Isle in 1928, humans knew next to nothing about coral reefs. Their work sparked worldwide interest in these amazing places. People flocked to the reef in ever-increasing numbers, lured by its ethereal beauty, its seemingly inexhaustible productivity, and the promise of riches. But with the attention came damage. Some foresaw what would happen if things didn't change, predicted the damage that would occur if quarrying and oil production were allowed to go ahead. They raised the alarm and galvanised public opinion, and brave leaders made courageous choices based on the science as they understood it at the time, and based on what was right. This unique bequest from the past was simply too valuable, too precious and too unique to risk. The precautionary principle applied, as it must continue to today.

And because of these courageous people, the Yonges and Hedleys and Verons and Wrights and Hoegh-Guldbergs, and the politicians who have made the decisions and enacted the laws, we can still enjoy the beauty of this place. Even if we might never see it in person, we are all richer for knowing it is there, for having the option to visit it in the future if we so choose. What is still in the balance is to ensure that our children or theirs or theirs might also have this privilege.

~

Humans have always used stories to capture the past, understand the present, and imagine the future. Poems, films and novels can educate, warn and motivate. Perhaps never before in our history have we needed that power of imagination quite as much as we do now.

The famous writer and futurist HG Wells once said that 'human history becomes more and more a race between education and catastrophe'. That was back in 1919. The calamity of the First World War was finally over and millions lay dead. Many felt the so-called 'war to end all wars' had been entirely preventable and the sacrifice unjustified. The treaty that was to set the stage for the next war was being negotiated, and a terrible pandemic was sweeping the planet, killing millions more.

Wells's message was clear: the future is inherently uncertain, but it is only through education, the spread of knowledge and the advancement of science that we can hope to influence it for the better. We can't change the past, but we can learn from it. And we certainly can't avoid the future – it will come soon enough, with or without us. All we have is the present, that constant but strangely elusive *now* of our everyday existence. And we must use it to create the future we want.

Looking into the future, good or bad, is foremost an act of imagination. Who, in 1919 – other than maybe Wells himself – might have imagined that the world's population would have more than quadrupled by 2023, reaching over eight billion? Who could have predicted the loss of so much of our natural environment, the disappearance of species and the rapid changes to our planet's climate, driving more intense wildfires, drought and extreme weather of all kinds? Yonge and Veron both spoke about their youthful impressions of a Great Barrier Reef far too big and majestic ever to be threatened by human activity, in a time before either could have imagined the juggernaut that is human-made climate change. It would be like asking someone today to predict what the world will be like in 2123. It's just so far away.

And yet, writers and scientists have been imagining the future for a long time. The IPCC reports, six of them since 1990, have consistently warned what would happen if we stayed with business as usual and kept burning fossil fuels and clearing forests. If we had started to act three decades ago, the present would look very different. The conversation we are having now about the GBR would be very different. But Professor Hoegh-Guldberg's warnings back in 1999 were ignored. So were Tim Flannery's in 2005.[7] And so were Gwynne Dyer's, chillingly delivered in his seminal 2010 book *Climate Wars*, which envisaged global conflict driven by climate change.[8] Even those of the US Department of Defense, which identified climate change as a major 'threat multiplier' to world stability, have been quietly discounted.[9]

Some of us – many of us – read and understood these warnings, were even galvanised to personal action. But it has never been enough. The imaginers, those providing the warnings, have been dismissed as doom-merchants, alarmists. And so we have dithered and obfuscated, clinging to that warm illusion that things will be okay, that somehow everything will turn out alright. We have run, collectively, to the coward's refuge: wishful thinking.

If there was any doubt that the time for courage had arrived, in 2023 a new powerful El Niño cycle started building in the Pacific, releasing heat stored in the oceans. The northern hemisphere summer arrived with a vengeance. My sister-in-law and her husband in Canada were forced to evacuate their home as wildfires bore down on their property. They collected the few things they could not bear to lose and were moved to the other side of the river, where they were put up in a hotel at government expense. More than a thousand fires raged across the country during the hottest summer of the hottest year ever on record. Day after day, they and thousands like them waited to hear if their fire had been brought under control, and if their home had survived.

At the same time, coral reefs from the Caribbean to the Pacific were starting to bleach.[10] The world's oceans continued to accumulate heat, driving more extreme storms. Temperature records were being broken all over the world.[11] Devastating wildfires burned out of control in Hawaii and Greece and Canada, destroying property and killing untold millions of native animals. Both poles were showing signs of unprecedented melting, and glaciers around the world were retreating faster than ever. Researchers warned that ecosystems around the world, including reefs, were now at risk of collapsing much more quickly than previously supposed.[12]

Global inequality has been growing too, among nations, between rich and poor, between old and young. As wealth has become increasingly concentrated in fewer and older hands, the rich keep getting richer and the poor fall further and further behind. The more we have delayed action on climate change, the more we have pushed the costs of damage and eventual action onto younger people and future generations. Maybe that was the plan all along.

My sister-in-law's house was saved from the fire, but many others lost everything. In the southern hemisphere, we watched nervously as winter turned to spring. Would the El Niño bring the same terrible summer heat, fires and bleaching we had seen in the north? Climate change in 2023 was no longer some distant prediction, something that might affect our grandchildren. It was happening now. To us. The planet's climate was changing before our eyes, and the best available science told us unequivocally that it would get a lot worse if we didn't do something quickly.

Human beings have always looked to the future with a mixture of hope and trepidation. It's why we take out mortgages and buy insurance, why we have children and do our best to make sure they get a good education.

Planning for our own futures takes up a lot of our time and energy. We hope for the best and prepare for the worst, guided by experience and what we know of the world around us. But what if the future is beyond our experience?

There is an old Danish proverb that says, 'It is difficult to make predictions, especially about the future.' How will the future judge us, we who have lived in these critical decades for the reef and the world? I suppose in a way, they are all critical decades, down through history. But there are so many of us now, wielding such technological power, that the effect of our decisions is magnified many times over what it was even a couple of generations ago. In the end, perhaps our inaction has been the result of a collective failure of imagination. Unable to see just how bad it was going to get, immobilised by uncertainty and fear, tricked by the soothing voices of inaction, we lost our way.

But it doesn't have to be this way. There is still time to shape the future we want. All we need is the courage to act.

~

As you approach the frontline, you can feel the change. It's as if the air has thickened, taken on a quality you have never noticed before. You can feel it flowing viscous and fluid-like into your lungs, and leaving again in anxious exhalation. Perhaps it is the columns of smoke you can see in the distance, grey against the deep winter blue, or the knowledge that you are now in artillery range.

Everyone you are with is aware of this line being crossed. Out of range, and then, with no particular milestone to mark the crossing, in. Everything has changed and yet nothing. Moving across into this territory of sacrifice has been an act of conscious will. All the people you are with know this. Everyone is scared. They know that whatever happens, they will be irrevocably changed.

Being here is a decision, perhaps taken long ago, maybe based on principle or just the need to do something in the face of evil. Whatever their personal reasons for being here, the people in the vehicle with you are the living definition of courage. Something must be done, and so they will be the ones to do it, regardless of the sacrifice called for.

There is a notion seemingly ingrained in Western thought at the moment that the great general public – the vast heaving mass of voting and non-voting humanity – cannot tolerate sacrifice for the greater good. We the public are so weak and irresolute that we must be treated like cosseted children, protected from reality and fed a steady diet of distracting trivia lest we be irretrievably damaged. For decades now, scientists have been warned not to present the facts in their full stark reality, lest the population be immobilised and demotivated by fear.

And so we have kept the message benign and hopeful, underplayed the consequences and overplayed the progress being made. The thinking was that if faced with the immensity of the challenges before us, and the scale of the sacrifice needed, people would feel so disempowered that they would simply give up and retreat into their comfortable world of Netflix and beer and pizza.

For most of us who are lucky enough to live in developed countries, the sacrifices called for to deal with climate change once and for all and to protect the reef won't be a matter of life or death. It is going to mean spending more of our tax dollars on energy transition and climate adaptation projects like RRAP. For some, especially high-income earners, it will mean higher taxes. For most of us, there will have to be restrictions on what we can do and how we can do it, some of which might have to last for decades. Transitioning the world's energy system will cost trillions of dollars. Adapting our homes, cities and infrastructure will cost trillions more. Because we have wasted the last four decades, everything will have to be done much more quickly, at higher cost, and at the same time. This will drive up prices, as similar projects compete for finite resources and know-how. It's a great time to get into engineering or project management.

How disdainful those who say we cannot hear this message of sacrifice and hard work must be of their fellow citizens. We should all be insulted to be thought of as weak and fragile cowards, unable to cope with the truth. If we have acted this way, it is only because we have been trained to be like this, passive and compliant. The climate deniers would have it that *they* are the ones challenging the system, daring to voice a contrary view. But this is just another example of the card trick they have been playing on the public these last few decades. The deniers are the voice of the status quo. They are the voice of the current economics, a rear-guard action for the fossil fuel industry. And it has worked. Despite all the talk and pledges, global emissions of the gases that are driving climate change actually rose in 2023, reaching an all-time record.[13] At the time of writing, there is still no evidence that the curve has even started to flatten.[14] The 2023 Conference of the Parties to the UN Framework Convention on Climate Change, COP28, held ironically in Dubai, in the petrostate of the United Arab Emirates, and chaired by an ex-oil industry executive, again failed to deliver the brave, ambitious changes required to avoid catastrophe. And so temperatures keep rising, and the reef's future remains in peril.

I have seen courage in the face of fear. We have all seen examples of bravery, big and small. This is what we need now – real courage. Not fake commitments to action for political gain, or empty treaties we do not intend to honour, or bogus commitments to emissions reduction that we have no intention or ability to honour. We don't need soothing rhetoric about growth and ever-increasing wealth and no-pain solutions. No more pretending that we are somehow going to get through this without sacrifice. Fighting climate change to an uneasy truce and saving the Great Barrier Reef and the world's other critical ecosystems is going to take real effort, money, resources and a huge change in expectations. And this burden can only fall to people alive now. If we leave it to future generations, it will be too late. The time is now, and it's us, all of us, who have to carry the burden.

It all sounds daunting and scary. But the good news is that we already have all of the technology and know-how we need to turn things around. We are rich enough – the cost of the Iraq and Afghanistan wars alone would have been enough to do most of the job a decade ago. The only thing we still need to find is the will and the courage to act. The benefits, eventually, will be massive. A healed world with stable ecosystems providing food and livelihoods and wonder for us all, for many generations to come. A new clean energy system where most of the fuel is free and accessible to anyone. An end to lopsided distribution of wealth and opportunity. A new age of prosperity for all.

In his novel *For Whom the Bell Tolls*, Ernest Hemingway wrote, 'The world is a fine place and worth the fighting for.' If we work hard now, it can stay that way for a long time to come. But there can be no more procrastination. It is time to get out of bed, each of us, leave the cosy warmth of our safe places, put on our body armour, take a deep breath and get going. Right now, there are battles that need to be fought.

17

Taking Action

In February of 2024, after the hottest January yet recorded on planet Earth,[1] and with February also about to set temperature records,[2] I caught up with Anna Marsden, managing director of the Great Barrier Reef Foundation.* By then, things were going well at the foundation, and they were riding a wave of fundraising success, catalysing projects across the reef, pulling in major donors from across the world wanting to help save the GBR, and by extension the world's reefs. 'This will be an amazing year,' she told me, irrepressibly optimistic as always. 'We have a big opportunity. Coral reefs are among the first ecosystems on the planet failing as a result of climate change. They are symbols of what is to come. We may have more bleaching in the GBR this year. We are at a crossroads. To save the reef, we have to be ambitious. It's up to us to show that it can be done.'

Darren Kindleysides, CEO of the Australian Marine Conservation Society, who have been advocating for the reef since their involvement in the earliest efforts to prevent mining and oil production on the GBR

---

* In an unsettling parallel, articles from 2016, another El Niño year, had eerily similar headlines, like this one in *The Observer*: 'February was the warmest month in recorded history, climate experts say'.

in the 1960s, agreed. 'Maintaining hope is the key,' he told me in early 2024. 'Over the years, the reef has faced many different challenges. But each time the warning went out, and action was taken. We have made so much progress. Climate change is an enormous threat, but if we work together, listen to the science, we can swing things around.'

By March of 2024, signs of mass bleaching were once again being reported across the Great Barrier Reef. Cyclone Kirrily, which made landfall near Townsville in late January, had brought some cooling, but not enough. Even the reefs on Lord Howe Island much further south started to bleach.[3] AIMS and GBRF began their aerial surveys, and by early March they were reporting significant and widespread coral bleaching in the southern part of the GBR, and low to moderate bleaching in the north and central areas.

In late March, I caught up with Dr Neal Cantin, the AIMS scientist who had shown me the coral cores back in 2017 during my first week at the institute. He had just completed the last of the season's aerial surveys of the reef, having covered the entire length of the GBR over the previous weeks. He was unequivocal about what he was seeing: the reef was bleaching again, and in places severely. Coral was starting to die, and climate change was the cause. 'It's hard watching this latest bleaching,' he told me. 'It wasn't supposed to be happening this frequently until 2035. It's scary because we are only a third of the way into the warming trend. If we're seeing this with 1.1 degrees of warming, I hate to think what it will be like at 3 degrees.'

And worst of all, heat was continuing to accumulate. NOAA's Coral Reef Watch warned that accumulated heat stress in the southern GBR was at an all-time high, surpassing even 2016,[4] and that the world was on the brink of another global mass coral bleaching.[5] If April continued hot, bleaching could lead to more coral death. We won't know how extensive the damage is until after AIMS completes its long-term monitoring later in the year. By the time you are reading this, we will know the extent of the damage.

Whatever happens this southern hemisphere summer, it is now almost certain that one or more catastrophic bleaching events will hit the Great Barrier Reef some time in the next decade. It could well be sooner rather than later. According to Dr Line Bay, 'It is more a matter of when than if.' Many reef scientists share her view. It is also highly likely that these events will cause coral death at a scale not seen since 2016 and 2017. With each successive die-off, it will get harder for the reef to bounce back. Each hit will lower the in-built resilience of the reef, to the point where some reefs will simply die and not recover. We have already seen that happen. Some northern GBR reefs hit hard in 2016–17 have not recovered, and lie still and silent under the blue surface of the water, lifeless testaments to human folly. Nothing we can do will change this. We have waited too long, wasted too much time. This will be the price of our cowardice.

It is a hard thing to come to terms with. It's just so damn sad. But there is still hope. If we get going right now, and apply ourselves fully to the task, much of the Great Barrier Reef can be saved. It may look a little different than the best of it does today, but it will still be a marvel worth travelling to see, worth dreaming about, worth being proud of.

And the really good news is that if we get to keep the reef, so much more that we treasure and depend on will also be made safe. Forests. Wetlands. Coastlines. Favourite holiday spots. Clean water. Shacks by the sea. Songlines and stories. Black cockatoos. Wonder. Strangers in faraway places. Peace. Dreaming.

Here's how.

Individual Action

Inform Yourself

Despite the clutter and noise, there is plenty of good information out there about what is going on. Trust the websites of the Great Barrier Reef Marine Park Authority and AIMS. The information presented there is the best available, and completely trustworthy. Try Google Scholar and have a look at what the peer-reviewed literature is actually saying. And trust it – it's the best we have. It takes a little more effort to access, but the abstracts of most papers will usually give you a pretty good idea of what the authors are getting at. Most importantly, don't fall prey to the honeyed, simplistic voices of those who would have you believe that everything is just fine and that you don't need to change. If it sounds too good to be true, it's because it is.

Vote

If you are lucky enough to live in a democratic country, exercise your franchise. The really big decisions that will result in collective action can only be made by governments – local, state and national. Recent work by the CSIRO shows that creating the future we want will be 70 per cent the result of collective action, and 30 per cent due to individual action.[6] So vote for candidates and parties who understand the issues and have the courage to make the changes we need, and who respect you enough as a voter to give you the straight truth on what is needed and what you will have to contribute to get there. Vote for candidates who will continue aggressive action to reduce greenhouse gas emissions, and who will enhance protection of the Great Barrier Reef and the country's other coral reef systems and help them adapt to warming oceans. Support parties that will stop the clearing of what's left of our native vegetation, give Indigenous people more control over their traditional lands and sea country, promote sustainable agriculture, and create greener, more liveable cities.

Make Your Voice Heard

Talk to friends and neighbours about what is going on. Write to your elected representatives and tell them what you want. Call out the bullshit when you hear it. Tell the companies you deal with that you will not buy their products or services unless they take strong action to reduce emissions, cut back on waste, take responsibility for the safe disposal of their products once they have reached end-of-life, and make their operations more sustainable. Imagine the world you want, and describe it to people.

Make Changes in Your Own Life

Go for low-emissions transport. Make your home more energy efficient. Quit plastic. Install solar panels and batteries and go electric. Insist on renewable energy, and buy it for your home. Plant native trees and shrubs, and keep the trees you already have. Offset your carbon emissions with certified biodiverse plantings. Help out in the community. Ride your bike or walk those shorter journeys. Eat organic, local food, and cut back on meat. All the things you already know about. But whatever you do, don't be fooled by the big companies who would have you believe that the responsibility for making change lies *exclusively* with you. For years, big companies have been pushing this line of individual action as a way of deflecting responsibility away from themselves. Do your part at home and in your own life, but make sure that big companies and institutions do theirs too.

Stay Positive and Realistic

It's too easy to succumb to fear and negativity. Know that we live in a time when everything matters. Every single thing we do from here on that moves the dial in the right direction will make a difference. Yes, things are looking shaky. But without action, they will get unimaginably worse. All the little things matter, and the big things matter more. Remember, the naysayers and the deniers *want* you to lose heart and give up. They want and need your silence, your inaction. That way, they get to keep things the way they are for a little bit longer, squeeze a little bit more profit out of the system. Stay strong, stay brave, and lean on others who want the same things you want.

Give Your Time and Money to Things You Care About

If you want to help save the Great Barrier Reef, or any other places you treasure, there are a lot of things you can do. Donate to charities that direct funds to helping improve conditions on the reef. Support NGOs that are doing the hard campaigning for the reef, organisations such as the Australian Marine Conservation Society and the World Wide Fund for Nature. You may not agree with everything they do, but they have been fighting to protect the GBR for decades, and they have made a difference. Despite the unfair bashing it's had, the Great Barrier Reef Foundation does great work and supports dozens of far-reaching and highly impactful projects. Other GBR charities and NGOs such as Citizens of the Great Barrier Reef and the Reef Restoration Foundation are filling important niches in the effort. There are loads of opportunities to volunteer on citizen science projects, turtle conservation efforts, beach clean-ups and the like. And best of all, just visiting the reef helps it. Tourism operators on the GBR are the most sustainable on the planet, and going there shows politicians that you care and that you want it protected.

Government and Organisational Action

Support Positive Action on the GBR

Governments and other organisations have already put a lot of money into programs designed to help the reef. Given the current situation, this needs to be sustained and increased. Efforts to improve water quality and control crown-of-thorns starfish all help to maintain the reef's fundamental resilience, and need to continue. Research into ways of helping the reef adapt to warming conditions needs to accelerate. Unfortunately, these new techniques will be needed soon enough. The GBRMPA needs to be supported to continue to do its job of regulating the various uses on the reef, and needs to be ready to approve and support interventions to help the reef adapt to climate change when they come. And critically, greenhouse gas emissions, both here and abroad, need to start coming down, fast.

Change the Tax Regime

Taxes raise revenue, but they are also used to discourage unwanted behaviours. Think smoking and booze. Currently, our system largely taxes good things: value added, income, profit, capital gains. We should be taxing bads, not goods. So, change the system to tax all the things we don't want: carbon emissions, waste, environmental damage of all things, erosion of natural capital, fossil fuels, packaging, health damage. And stop subsidising damaging activities. With all the extra revenue generated, we could reduce taxes on income and wealth generation. The transformation would be widespread and swift. People would have more money to spend, and price signals would drive expenditure away from damaging products and services and towards safe ones. The market would work quickly to reallocate resources away from damaging enterprises and into projects designed to accelerate the transition to the future we need.

Make Better Decisions

For big projects and major policy decisions, start including all of the environmental, social and economic costs and benefits in the assessment of a range of options, and consider cumulative effects. Currently, very few decisions are made this way. At present, the environmental and social costs of a project are rarely estimated. If damage is anticipated, it is usually expressed only in qualitative terms, or in units other than money. By expressing damage in monetary terms as well, decision-makers can compare the benefits of various options against the overall damage to society that they will entail.[7] Legislating a requirement for triple-bottom-line assessment of project options would rapidly move decision-makers away from damaging projects and towards beneficial ones.

Adopt a New Measure of Success

Currently, Australia and most other countries use Gross Domestic Product, GDP – the total value of goods and services produced – as the key and virtually only measure of economic success. If the rate of growth of GDP drops even for a few months, politicians and central bankers start getting nervous, and the fearful words 'recession' or even, God help us, 'depression' start being whispered. In contrast, when GDP grows, everyone lines up to take credit for it, even if, as is currently the case in Australia, GDP growth is mostly due to a growing population and not an underlying improvement in productivity.

But GDP is a hugely flawed way of measuring success in society.[8] It does not include many things that are good for society, such as the vast amounts of unpaid labour contributed by volunteers and carers. Nor does it value the services provided by nature. A forest is not valued for the life-giving biodiversity it harbours, the water purification it provides or the oxygen it pumps into the atmosphere. Only when cut down and converted to lumber or wood chips is the forest deemed to have value and therefore included in GDP.

On the other side of the ledger, GDP does *not* include the value of any of the damage we do to society or the environment as a result of our activities. So when we change the climate by pumping billions of tons of carbon into the atmosphere, the value of the resulting damage is not included in GDP (as a negative), but the value of the fuel burned and the goods produced is. If an oil well blows out in the Gulf of Mexico, none of the damage to the environment is included as a negative in the GDP calculation, but the costs of the clean-up are counted as a positive. Oil spills are good for GDP. The damage to the environment caused by climate change is not recorded as a negative to GDP. But anything we do to cope with its effects, perversely, is good for GDP. If we spend money to help our cities adapt to rising sea levels, to recover after floods and fires and other extreme weather events, to help the coral reefs of the world adapt to warming oceans, it is counted as a positive in the GDP ledger. At least in the short term, climate change is good for GDP.

Collectively, all of these distortions of reality drive us, again and again, through the obsessive application of the concept of GDP, to decisions that erode and pollute the natural world, make us all less well-off and ignore or stigmatise people who do unpaid work.

But as Nicholas Stern showed in *The Economics of Climate Change*, the clock is ticking, even for GDP. As climate change starts to bite hard, the damage will become so great that economic activity everywhere will slow. Many alternative measures of success have been proposed, including Net Economic Welfare, which adjusts GDP for hidden environmental and social costs and benefits. Some countries have already introduced such measurements. We need to get rid of or modify GDP now, while we still have time, and adopt new measures of success.

Demonstrate Leadership

People at all levels of society need to show leadership if we are to overcome the challenges we are now facing. Whether you are an elected representative, a senior executive, a professional architect or engineer or scientist, a mine worker, tradie, stay-at-home mum or dad, teenager, or school kid, leadership means doing what is right, not just what is expedient. It means telling the truth, and treating the people you lead with respect. Good leaders take responsibility, practise humility, park their egos and put the team's success first. They listen and are readily recalled to conciliation. They are disciplined and apply themselves to things that truly matter. They set clear goals and communicate them. Leaders embrace ambiguity and practise decisiveness in the face of uncertainty. They overcome fear in themselves and help others overcome theirs, and so face life's gales with determination and courage. To win this fight, we are all going to need to show genuine leadership.

In the end, this is a lot more than a war for a reef, even if it is the biggest and most magnificent one on the planet. The reef is a symbol for our planet – a wondrous, diverse, but ultimately fragile organism set in a vast ocean. And yet, for a long time, its sheer size made people think it was invulnerable. A century ago, it was inconceivable that our puny efforts could ever degrade something so wonderfully vast.

But illusions can be dangerous things. The reef, and the planet that hosts it, are in deep trouble. Don't let the transcendent beauty of that blue surface fool you. Underneath, destructive forces are at work, unleashed by our own hand, and they are now close to spinning out of control. But there is still time to avert the worst. Not much time, but enough.

There are many alternative futures that might yet unfold. We need to find the courage to shape the future we want, before the future shapes us. Just imagine.

Acknowledgements

As always, no book comes about without the efforts of people who only get thanked in an acknowledgements section like this.

Thanks first to my family and friends, the special people in my life who every day inspire and encourage me to write, to tell the stories that need to be told, and who give me the courage to keep going, despite the perils and challenges.

Thanks to Martin and Armelle at Affirm Press for their incredibly valuable help and guidance. Without them, this book would never have come about.

Deepest thanks to all of the fantastic people at AIMS for a wonderful six years, and for your continued friendship and support. Keep going, you guys: keep being the best, keep being uncompromising about the quality of the science you do.

Thanks to everyone who agreed to be interviewed for this book, and who helped me make sure I got the story right.

And above all, thanks to you, the reader. If you have got this far and are reading this, then I hope the journey has been an enjoyable one, and has in some way informed and inspired. If it has, I look forward to seeing you out there one day snorkelling the reef, at a rally to protect a forest, volunteering in your local community, or on Capital Hill making your voice heard.

References

Chapter 1: Into the Fray

1   ABC News, 'Pauline Hanson visits healthy reef to dispute effects of climate change', 25 November 2016, abc.net.au/news/2016-11-25/pauline-hanson-visits-the-great-barrier-reef-climate-change/8059142.

2   'Climate', Pauline Hanson's One Nation, onenation.org.au/climate, accessed 7 March 2024; *SBS News*, 'Hanson denies humans behind climate change, blames "fearmongering"', 23 April 2019, sbs.com.au/news/article/hanson-denies-humans-behind-climate-change-blames-fearmongering/2y23ihwxy.

3   Griffith, Hywel, 'Great Barrier Reef suffered worst bleaching on record in 2016, report finds', BBC News, 28 November 2016, bbc.com/news/world-australia-38127320.

4   Jacobsen, Rowan, 'Obituary: Great Barrier Reef (25 Million BC–...)', *Outside Magazine*, 11 October 2016, outsideonline.com/outdoor-adventure/environment/obituary-great-barrier-reef-25-million-bc-2016/.

5   Deloitte Access Economics, *AIMS Index of Marine Industry 2020*, report for the Australian Institute of Marine Science, 2020, aims.gov.au/sites/default/files/2021-07/The%20AIMS%20Index%20of%20Marine%20Industry_final_21Jan2021_web.pdf.

6   Murphy, Katharine, 'Scott Morrison brings coal to question time: what fresh idiocy is this?', *The Guardian*, 7 February 2017, theguardian.com/australia-news/2017/feb/09/scott-morrison-brings-coal-to-question-time-what-fresh-idiocy-is-this.

7   Wilkinson, Clive, and David Souter (eds), *Status of Caribbean Coral Reefs after Bleaching and Hurricanes in 2005*, Global Coral Reef Monitoring Network, Townsville, 2008.

Chapter 2: Why Care?

1   Bartley, Rebecca, et al., 'Chapter Two: Sources of sediment, nutrients, pesticides and other pollutants to the Great Barrier Reef', *2017 Scientific Consensus Statement: Land use impacts on Great Barrier Reef water quality and ecosystem condition*, Reef Water Quality Protection Plan Secretariat, Brisbane, 2017.

2   Souter, David, et al. (eds), *Status of Coral Reefs of the World, 2020*, International Coral Reef Initiative, ICRI, Paris, 2020.

3   Wilkinson, Clive, et al., 'Tropical and Sub-Tropical Coral Reefs' in Lorna Inniss and Alan Simcock (eds), *The First Global Integrated Marine Assessment: World Ocean Assessment I*, 2016, http://www.un.org/depts/los/global_reporting/WOA_RPROC/Chapter_43.pdf.

4   Deloitte Access Economics, *At what price? The economic, social and icon value of the Great Barrier Reef*, report for the Great Barrier Reef Foundation, 2017, deloitte.com/content/dam/assets-zone1/au/en/docs/services/economics/deloitte-au-economics-great-barrier-reef-230617.pdf.

5   Ibid.

6   Souter et al., *Status of Coral Reefs of the World: 2020*.

7   Deloitte Access Economics, *At what price?*.

8   Souter et al., *Status of Coral Reefs of the World: 2020*.

9   Deloitte Access Economics, *At what price?*

Chapter 3: Yonge and the Early Days of Exploration

1   Fielding, Trisha, *Expedition to the Great Barrier Reef: The story of a ground-breaking scientific mission to Low Isles, Queensland in 1928–1929, and an overview of its legacies*, James Cook University, Townsville, 2023.

2   Harrison, Peter, et al., 1984, 'Mass Spawning in Tropical Coral Reefs', *Science*, vol. 223, no. 4641, doi.org/10.1126/science.223.4641.1186.

3   Yonge, C. Maurice, *A Year on the Great Barrier Reef: The Story of Corals and of the Greatest of their Creations*, Putnam, London, 1930.

4   Rowland, Michael, 'Myths and Non-Myths: 'Frontier "Massacres" in Australian History – The Woppaburra of the Keppel Islands', *Journal of Australian Studies*, vol. 28, no. 81, 2004, doi.org/10.1080/14443050409387934.

5   Ibid.

6   Ibid.

7   Muir, Bob, Personal communication (interview), May 2022.

Chapter 4: Truth and Denial

1   Queensland Government Department of Agriculture and Fisheries, 'Primary Industries Data', 2023, daf.qld.gov.au/news-media/campaigns/data-farm/primary-industries.

2   Waterhouse, Jane et al., *2017 Scientific Consensus Statement: Land Use Impacts on Great Barrier Reef Water Quality and Ecosystem Condition*, Department of the Environment and Energy, Brisbane, 2017, reefplan.qld.gov.au/__data/assets/pdf_file/0029/45992/2017-scientific-consensus-statement-summary.pdf.

3   Institute of Public Affairs, 'Dr Peter Ridd discussing the health of the Great Barrier Reef on Sky News Australia – 16 April 2023', 21 April 2023, youtube.com/watch?v=Y3Ph9QwW5-E.

4   Sky News Australia, 'Intellectually "bankrupt" scientists are running down the Great Barrier Reef', 6 August 2023, youtube.com/watch?v=wFthjWjiVT0.

5   Ridd, Peter, 'The Great Great Barrier Reef Swindle', *On Line Opinion*, 19 July 2007, onlineopinion.com.au/view.asp?article=6134&page=0.

6   Wilkinson, Marian, *The Carbon Club*, Allen & Unwin, Sydney, 2020.

7   Ibid.

8   Roskam, John, Chris Berg and James Paterson, 'Be Like Gough: 75 Radical Ideas to Transform Australia', Institute of Public Affairs, 5 August 2012, ipa.org.au/ipa-review-articles/be-like-gough-75-radical-ideas-to-transform-australia.

9   Ibid.

10 Ridd, Peter, 'The Greens: illogical and treacherous', *On Line Opinion*, 12 May 2008, onlineopinion.com.au/view.asp?article=7352&page=0.

11 Readfern, Graham, 'Were Historical Pictures of Great Barrier Reef Degradation Really Misused, as The Australian Newspaper Claimed?', *DeSmog*, 16 June 2016, desmog.com/2016/06/16/were-historical-pictures-great-barrier-reef-degradation-really-misused-australian-newspaper-claimed/.

12 ——, 'University fires controversial marine scientist for alleged conduct breaches', *The Guardian*, 21 May 2018, theguardian.com/environment/2018/may/21/university-fires-controversial-marine-scientist-for-alleged-conduct-breaches.

13 Commonwealth, *Parliamentary Debates*, Senate, 28 November 2016, 3389, (Malcolm Roberts).

14 Ibid.

15 Ibid.

16 Ibid.

17   Steele, Jim, *Landscapes and Cycles: An Environmentalist's Journey to Climate Skepticism*, 2013, http://landscapesandcycles.net. /coral-bleaching-debate.html.

18   Barry, Paul (presenter), 'Muddying the waters on the Great Barrier Reef', *Media Watch*, ABC, episode transcript, 18 July 2016, abc.net.au/mediawatch/episodes/muddying-the-waters-on-the-great-barrier-reef/9972936.

19   Desikan, Anita, et al., 2023, 'An equity and environmental justice assessment of anti-science actions during the Trump administration', *Journal of Public Health Policy*, vol. 44, doi.org/10.1057/s41271-023-00460-3.

20   Ridd, Peter, 'Science or Silence? My battle to question doomsayers about the Great Barrier Reef', Fox News Opinion, 8 February 2018, foxnews.com/opinion/science-or-silence-my-battle-to-question-doomsayers-about-the-great-barrier-reef

21   Horbach, Serge, and Willem Halffman, 2018, 'The Changing Forms and Expectations of Peer Review', *Research Integrity and Peer Review*, vol. 3, no. 8, doi. org/10.1186/s41073-018-0051-5.

22   Ridd, 'Science or Silence?'

23   Schaffelke, Britta, et al., 2018, 'Support for improved quality control but misplaced criticism of GBR science. Reply to viewpoint "The need for a formalised system of Quality Control for environmental policy-science" by P. Larcombe and P. Ridd (Marine Pollution Bulletin 126: 449–461, 2018)', *Marine Pollution Bulletin*, vol. 129, no. 1, doi.org/10.1016/j.marpolbul.2018.02.054.

24   Ridd, Peter, 'Reef "Doomsayers" Show Breakdown In Our Science Institutions', *Institute of Public Affairs*, 10 August 2023, ipa.org.au/publications-ipa/opinion/reef-doomsayers-show-breakdown-in-our-science-institutions.

25   Environment and Communications References Committee, *Inquiry into the Great Barrier Reef 2050 Partnership Program*, report, Department of the Senate, Canberra, 2019, aph.gov.au/Parliamentary_Business/Committees/Senate/Environment_and_Communications/GBRPartnershipProgram/Report.

Chapter 5: Visions for the Reef

1   Intergovernmental Panel on Climate Change, *Special Report on Emissions Scenarios: A Special Report of IPCC Working Group III*, Cambridge University Press, 2000.

2   American Petroleum Institute, *Global Climate Science Communications Action Plan*, 1998.

3   Banfield, Edmund, 1908, 'Dunk Island: Its General Characteristics', *Queensland Geographical Journal*, vol. 23.

4   Bowen, James, and Margarita Bowen, *The Great Barrier Reef: History, Science, Heritage*, Cambridge University Press, Melbourne, 2002.

5   Richards, Henry Casselli, 'Problems of the Great Barrier Reef', *Queensland Geographical Journal*, vol. 36–37, 1922, espace.library.uq.edu.au/view/UQ:382599.

6   Saville-Kent, William, *The Great Barrier Reef of Australia: Its Products and Potentialities*, WH Allen & Co., London, 1893.

7   Mackay, John (commissioner) et al., *Report of the Royal Commission Appointed to Inquire into the Working of the Pearl-shell and Beche-de-mer Industries ...*, Government Printer, Brisbane, 1908.

8   Bowen, *The Great Barrier Reef*.

9   *The Brisbane Courier*, 'Scientists Charmed', 22 September 1923, trove.nla.gov.au/newspaper/article/20635628.

10   Bowen, *The Great Barrier Reef*.

Chapter 6: The Fight for Funding

1   Karp, Paul, 'Head of reef foundation says $444m grant was "complete surprise"', *The Guardian*, 7 August 2018, theguardian.com/environment/2018/aug/07/head-of-reef-foundation-says-444m-grant-was-complete-surprise

2   Cox, Lisa, 'PM personally approved $443m grant for tiny Barrier Reef foundation' *The Guardian*, 31 July 2018, theguardian.com/australia-news/2018/jul/30/malcolm-turnbull-present-when-443-million-dollars-offered-to-small-group-without-tender-inquiry-hears

3   Environment and Communications References Committee, *Inquiry into the Great Barrier Reef 2050 Partnership Program*.

4   Ibid.

5   Ibid.

6   Ibid.

7   Ibid.

8   Ibid.

9   Baird, Lucas, 'Auditor-General finds $443m Great Barrier Reef Foundation passed tests', *Australian Financial Review*, 16 January 2019, afr.com/politics/auditorgeneral-finds-443m-great-barrier-reef-foundation-passed-tests-20190115-h1a30u.

10   Environment and Communications References Committee, *Inquiry into the Great Barrier Reef 2050 Partnership Program*.

Chapter 7: The Threat of Oil

1 Wright, Judith, *The Coral Battleground*, Nelson, Melbourne, 1977.

2 Lloyd, Rohan, *Saving the Reef*, University of Queensland Press, Brisbane, 2022.

3 Wright, *The Coral Battleground*.

4 Ibid.

5 Bell, Bethan and Mario Cacciottolo, 'Torrey Canyon oil spill: The day the sea turned black', BBC News, 17 March 2017, bbc.com/news/uk-england-39223308

6 Lloyd, *Saving the Reef*.

7 Bowen, *The Great Barrier Reef*.

8 Clarke, Keith, and Jeffrey Hemphill, 2002, 'The Santa Barbara Oil Spill: A Retrospective', in Darrick Danta (ed), *Yearbook of the Association of Pacific Coast Geographers*, University of Hawai'i Press, vol. 64, doi.org/10.1353/pcg.2002.0014.

9 Bowen, *The Great Barrier Reef*.

10 Australian Conservation Foundation, *The Future of the Great Barrier Reef*, proceedings of symposium, 3 May 1969.

11 Wright, *The Coral Battleground*.

12 Queensland, Parliamentary Debates, Legislative Assembly, 19 August 1969 (Ronald Camm, Minister for Mines, Main Roads and Electricity).

13 Bowen, *The Great Barrier Reef*.

14 'Forcing the Public to Direct Action', *The Australian*, 7 January 1970.

15 Lloyd, *Saving the Reef*.

Chapter 8: Biological and Geological Time

1 *Reef Water Quality Report Card 2017 and 2018*, Department of Environment, Science and Innovation, Queensland Government, 2019, reefplan.qld.gov.au/tracking-progress/reef-report-card/2017-2018 .

2 Great Barrier Reef Marine Park Authority, Australian Government, 2019, gbrmpa.gov.au/our-work/outlook-report-2019.

3 Limpus, Colin, *A Biological Review of Australian Marine Turtles*, Queensland Government, 2007.

4 Bock, Ellie et al., 2021, 'Safeguarding our sacred islands: Traditional Owner-led Sea Country governance, planning and management in Australia', *Pacific Conservation Biology*, vol. 28, no. 4, doi.org/10.1071/PC21013.

5 Pascoe, Bruce, *Dark Emu: Aboriginal Australia and the Birth of Agriculture*, Magabala Books, Broome, 2018.

6   *World Register of Marine Species*, Flanders Marine Institute, marinespecies.org/, accessed January 2021.

7   Food and Agriculture Organization of the United Nations, *The State of World Fisheries and Aquaculture 2014*, Rome, 2014, fao.org/family-farming/detail/en/c/286731/.

Chapter 9: Tragedies of the Commons

1   Hardin, Garrett, 1968, 'The Tragedy of the Commons', *Science*, vol. 162, no. 3859, doi.org/10.1126/science.162.3859.1243.

2   Hardisty, Paul, *Environmental and Economic Sustainability*, CRC Press, New York, 2010.

3   'Whale Oil Line Repaired', *The West Australian*, 28 September 1953, trove.nla.gov.au/newspaper/article/52931086/3798875.

4   'Whale Oil from Albany', *Daily Commercial News and Shipping List*, 6 October 1954, trove.nla.gov.au/newspaper/article/163997214.

5   Department of Climate Change, Energy, the Environment and Water, 'Whaling', 2021, dcceew.gov.au/environment/marine/marine-species/cetaceans/whaling.

6   Harrison, Peter, and John Woinarksi, 'Recovery of Australian subpopulations of humpback whale *Megaptera novaeangliae*', in Garnett, Stephen et al., *Recovering Australian Threatened Species*, 2018, CSIRO Publishing, Melbourne, 2018.

7   Australian Antarctic Program, 'Humpback Whale', 2022, antarctica.gov.au/about-antarctica/animals/whales/humpback-whale/.

8   'Whaling profit trebled', *The Canberra Times*, 12 January 1971, trove.nla.gov.au/newspaper/article/110448505.

9   Higgins, Jenny, '19th Century Cod Fisheries', 2007, *Heritage Newfoundland and Labrador*, heritage.nf.ca/articles/economy/19th-century-cod.php

10   Myers, Ransom, Jeffrey Hutchings and Nicholas Barrowman, 1997, 'Why Do Fish Stocks Collapse? The Example of Cod in Atlantic Canada', *Ecological Applications*, vol. 7, no. 1, doi.org/10.2307/2269409.

11   Hardisty, *Environmental and Economic Sustainability*.

12   Myers et al., 'Why Do Fish Stocks Collapse?'

13   Rose, George, and Sherrylynn Rowe, 2015, 'Northern cod comeback', *Canadian Journal of Fisheries and Aquatic Sciences*, vol. 72, no. 12, doi.org/10.1139/cjfas-2015-0346.

14   Pershing, Andrew, et al., 2015, 'Slow adaptation in the face of rapid warming

leads to collapse of the Gulf of Maine cod fishery', *Science*, vol. 350, no. 6262, doi. org/10.1126/science.aac9819.

15 Pinksy, Malin, et al., 2011, 'Unexpected Patterns of Fisheries Collapse in the World's Oceans', *Proceedings of the National Academy of Sciences*, vol. 108, no. 20, doi.org/10.1073/pnas.1015313108.

16 Patterson, Heather, Ashley Williams and David Mobsby, 'Southern Bluefin Tuna Fishery', *Fishery Status Reports 2020*, Australian Bureau of Agricultural and Resource Economics and Sciences, 2020.

17 Worldwatch Institute, *State of the World 1998: A Worldwatch Institute Report on Progress toward a Sustainable Society*, WW Norton, New York, 1998.

18 Patterson, Heather, and Michael Dylewski, 'Chapter 22 – Southern Bluefin Tuna Fishery', *Fishery Status Reports 2023*, Australian Bureau of Agricultural and Resource Economics and Sciences, 2023.

19 Hillary, Rich, 'Australian endangered species: Southern Bluefin Tuna', *The Conversation*, 17 January 2013, theconversation.com/australian-endangered-species-southern-bluefin-tuna-11636.

20 Patterson & Dylewski, 'Chapter 22 – Southern Bluefin Tuna Fishery'.

21 Stiglitz, Joseph, *The Price of Inequality*, Allen Lane, London, 2012.

22 Ibid.

23 Cheng, Lijing, et al., 2021, 'Upper Ocean Temperatures Hit Record High in 2020', *Advances in Atmospheric Sciences*, vol. 38, doi.org/10.1007/s00376-021-0447-x.

Chapter 10: Damage Control

1 Australian Cane Farmers Association, Submission to Rural and Regional Affairs and Transport References Committee, *Identification of leading practices in ensuring evidence-based regulation of farm practices that impact water quality outcomes in the Great Barrier Reef*, 8 November 2019.

2 Canegrowers, Submission to Rural and Regional Affairs and Transport References Committee, *Identification of leading practices in ensuring evidence-based regulation of farm practices that impact water quality outcomes in the Great Barrier Reef*, 2019.

3 Ibid.

4 Ridd, Peter, 'Dr Peter Ridd – Full presentation – Great Barrier Reef – August 2019', 18 August 2019, youtube.com/watch?v=itCQSSKwfHo.

5 De'ath, Glenn, Janice Lough, and Katharina Fabricius, 2009, 'Declining

Coral Calcification on the Great Barrier Reef', *Science,* vol. 323, no. 5910, doi. org/10.1126/science.1165283.

6 ——, 2013, 'Yes – Coral calcification rates have decreased in the last twenty-five years!', *Marine Geology,* vol. 346, doi.org/10.1016/j.margeo.2013.09.008.

7 Morton, Adam, and Ben Smee, 'Great Barrier Reef expert panel says Peter Ridd misrepresenting science', *The Guardian,* 28 August 2019, theguardian.com/ environment/2019/aug/28/great-barrier-reef-expert-panel-says-peter-ridd-misrepresenting-science.

8 AIMS, Submission to Rural and Regional Affairs and Transport References Committee, *Identification of leading practices in ensuring evidence-based regulation of farm practices that impact water quality outcomes in the Great Barrier Reef,* 8 November 2019.

9 Cook, Garry et al., 'Australia's Black Summer of fire was not normal – and we can prove it', *CSIRO,* 29 November 2021, csiro.au/en/news/all/articles/2021/ november/bushfires-linked-climate-change.

10 Hoegh-Guldberg, Ove, et al., 2019, 'The human imperative of stabilizing global climate change at 1.5°C', *Science,* vol. 365, no. 6459, doi.org/10.1126/science.aaw6974.

11 GBRMPA, *Great Barrier Reef Outlook Report 2019.*

12 Lloyd, Graham, 'Reef coral testing "flawed, needs fix"', *The Australian,* 2 January 2020, theaustralian.com.au/science/great-barrier-reef-coral-testing-flawed-needs-fix-says-peter-ridd/news-story/8c8432a1c8ee5ea13174356789cab310.

13 Ridd, Peter, 'Great Barrier Reef truth may be inconvenient but it is out there', *The Australian,* 2 January 2020, theaustralian.com.au/commentary/great-barrief-reef-truth-may-be-inconvenient-but-it-is-out-there/news-story/7584d10cde9ed9f7b9 00059be3118f81.

14 *The Australian,* 'Research Vital on Reef Coral', 3 January 2020, theaustralian.com. au/commentary/editorials/research-vital-on-reef-coral/news-story/47ddcd95ba3 7993623dda1b4fb0a741b.

15 Lloyd, Graham, 'Great Barrier Reef Foundation "spending millions on costs"', *The Australian,* 4 January 2020, theaustralian.com.au/nation/politics/ great-barrier-reef-foundation-spending-millions-on-costs/news-story/ badc755d7c4b9e91644575a5495e6d5e.

16 Hughes, Terry, et al., 2021, 'Emergent properties in the responses of tropical corals to recurrent climate extremes', *Current Biology,* vol. 31, no. 23, doi.org/10.1016/j. cub.2021.10.046.

17 McCallum, Hamish, 'New preliminary evidence suggests coronavirus jumped from animals to humans multiple times', *The Conversation*, 24 September 2021, theconversation.com/new-preliminary-evidence-suggests-coronavirus-jumped-from-animals-to-humans-multiple-times-168473.

18 World Health Organization, *WHO COVID-19 Dashboard*, data.who.int/dashboards/covid19/cases?n=c

Chapter 11: Layers of Defence

1 Wright, *The Coral Battleground*.

2 Committee of Inquiry into the National Estate and Robert Hope (chair), *National Estate : report of the Committee of Inquiry*, Government Printer, Canberra, 1974, nla.gov.au/nla.obj-1473866678.

3 Royal Commission into Exploratory and Production Drilling for Petroleum in the Area of the Great Barrier Reef, Wallace, Gordon and Queensland. Royal Commission into Exploratory and Production Drilling for Petroleum in the Area of the Great Barrier Reef. *Report*, Government Printer, Canberra, 1975, nla.gov.au/nla.obj-1918477690.

4 Lloyd, *Saving the Reef*.

5 Bjelke-Petersen to Whitlam, 25 November 1974, Queensland State Archives, SRS 1043. In Lloyd, *Saving the Reef*.

6 Commonwealth, *Parliamentary Debates*, House of Representatives, 22 May 1975, 2679 (Moss Cass, Minister for Environment).

7 Toyne, Phillip, *The Reluctant Nation*, ABC Books, Sydney, 1994.

8 UNESCO World Heritage Centre, 'The Criteria for Selection', whc.unesco.org/en/criteria/

9 GBRMPA, *Nomination of the Great Barrier Reef by the Commonwealth of Australia for Inclusion in the World Heritage List*, Townsville, 1981.

Chapter 12: Ambush

1 Commonwealth of Australia, *Official Committee Hansard*, Senate and Rural and Regional Affairs and Transport References Committee, 'Regulation of Farm Practices that Impact Water Quality Outcomes in the Great Barrier Reef', 27 July 2020.

2 Ibid.

3 Ibid.

4 Ibid.

5 Shteyman, Jacob, 'Senator wrong on greenhouse gas claim, say experts', AAP Factcheck, aap.com.au/factcheck/senator-wrong-on-greenhouse-gas-claim-say-experts/.

6 Arrhenius, S, 1896, 'On the Influence of Carbonic Acid in the Air upon the Temperature of the Ground', *Philosophical Magazine and Journal of Science*, series 5, vol. 41.

7 Commonwealth of Australia, 'Regulation of Farm Practices'.

8 Ibid.

9 Ibid.

10 Ibid.

11 Ibid.

12 Readfearn, Graham, 'Australia's science academy attacks "cherrypicking" of Great Barrier Reef research', *The Guardian*, 26 November 2019, theguardian.com/environment/2019/nov/26/australias-science-academy-attacks-cherrypicking-of-great-barrier-reef-research.

13 Chubb, Ian, Ove Hoegh-Guldberg and Geoff Garrett, response to written questions on notice, 28 August 2020 (received 3 September 2020). Quoted in Commonwealth of Australia, 'Regulation of Farm Practices'.

14 Commonwealth of Australia, 'Regulation of Farm Practices'.

15 Chubb, Ian, Geoff Garrett and Ove Heogh-Gulberg, 'At a Senate reef inquiry, we saw Australian politicians flaunt arrogance and ignorance', *The Guardian*, 18 September 2020, theguardian.com/commentisfree/2020/sep/18/at-a-senate-reef-inquiry-we-saw-australian-politicians-flaunt-arrogance-and-ignorance.

16 Commonwealth of Australia, 'Regulation of Farm Practices'.

17 Ibid.

18 Ibid.

19 Ibid.

20 Ibid.

Chapter 13: Warnings, Progress and Setbacks

1 Bowen, *The Great Barrier Reef*.

2 Veron, John, *Corals of the World*, AIMS, Townsville, 2000.

3 Harrison et al., 'Mass Spawning in Tropical Coral Reefs'.

4 Babcock, Russell and Andrew Heyward, 1986, 'Larval Development of Certain Gamete-Spawning Scleractinian Corals', *Coral Reefs*, vol. 5, doi.org/10.1007/BF00298178.

5   Bowen, *The Great Barrier Reef.*

6   Hoegh-Guldberg, Ove, and G Jason Smith, 1989, 'The effect of sudden changes in temperature, light and salinity on the population density and export of zooxanthellae from the reef corals *Stylophora pistillata* Esper and *Seriatopora hystrix* Dana', *Journal of Experimental Marine Ecology and Biology*, vol. 129, no. 3, doi. org/10.1016/0022-0981(89)90109-3.

7   IPCC, *Climate Change 2007: The Physical Science Basis*, Cambridge University Press, 2007.

8   Gribbin, John (ed), *The Breathing Planet: A New Scientist Guide*, Blackwell, New York, 1986.

9   Ibid.

10 Ibid.

11 NASA Earth Observatory, 'World of Change: Global Temperatures', earthobservatory.nasa.gov/world-of-change/global-temperatures

12 Wilkinson, Clive, 'The 1997–1998 Mass Bleaching Event Around the World', *Status of Coral Reefs of the World:1998*, AIMS, Townsville, 1998.

13 Fabricius, Katharina, 1999, 'Tissue loss and mortality in soft corals following mass-bleaching', *Coral Reefs*, vol. 18, doi.org/10.1007/s003380050153.

14 Jones, Ross, Ray Berkelmans and Jamie Oliver, 1997, 'Recurrent Bleaching of Corals at Magnetic Island (Australia) Relative to Seawater Temperature', *Marine Ecology Progress Series*, vol. 158, no. 1, doi.org/10.3354/meps158289.

15 AIMS, *Long-Term Monitoring of the Great Barrier Reef: Status Report 4, 2000*, AIMS, Cape Ferguson, 2000.

16 Hoegh-Guldberg, Ove, 'Climate Change, Coral Bleaching, and the Future of the World's Coral Reefs', *Marine and Freshwater Research*, vol. 50, 1999, doi. org/10.1071/MF99078.

17 Wilkinson, *The Carbon Club.*

18 Parliament of Australia, 'The Mabo Decision', aph.gov.au/Visit_Parliament/Art/Stories_and_Histories/The_Mabo_decision

19 GBRMPA, *The Great Barrier Reef: keeping it great. A 25 year strategic plan for the Great Barrier Reef Heritage Area 1994–2019*, GBRMPA, 1994.

20 March, Helene, *The Status of the Dugong in the Southern Great Barrier Reef Marine Park*, GBRMPA Research Publication 41, 1996.

21 Lucas, Percy et al., *The Outstanding Universal Value of the Great Barrier Reef World Heritage Area*, GBRMPA, 1997.

271

22 Haynes, D, and K Michalek-Wagner, 2000, 'Water Quality in the Great Barrier Reef World Heritage Area: Past Perspectives, Current Issues and New Research Directions', *Marine Pollution Bulletin*, vol. 41, no. 7, doi.org/10.1016/S0025-326X(00)00150-8.

23 Worldwatch Institute, *State of the World 1998*.

24 IPCC, *Climate Change 2001: The Scientific Basis*, Cambridge University Press, 2001.

25 IPCC, *Climate Change 2001: Impacts, Adaptation and Vulnerability*, Cambridge University Press, 2001.

26 GBRMPA, *Great Barrier Reef Marine Park Zoning Plan 2003*, GBRMPA, 2004.

27 IPCC, *Climate Change 2007: The Physical Science Basis*, Cambridge University Press, 2007.

28 Stern, Nicholas, *The Economics of Climate Change: The Stern Review*, Cambridge University Press, 2007.

29 Shiner, Meredith, 'Inhofe slams Gore on climate "hoax"', *Politico*, 15 March 2010, politico.com/story/2010/03/inhofe-slams-gore-on-climate-hoax-034443

30 Wilkinson, *The Carbon Club*.

31 World Bank, 'GDP (Current US$)', data.worldbank.org/indicator/NY.GDP.MKTP.CD

32 Bolt, Andrew, 'Global Warming – Just Settle Down and Wait for Proof. After All, Panicking is So Undignified', *Herald Sun*, 29 April 2009.

33 GBRMPA, *Great Barrier Reef Outlook Report 2009*, GBRMPA, Townsville, 2009, hdl.handle.net/11017/199.

34 Zethoven, Imogen, *The Last Decade: The World Heritage Committee and the Great Barrier Reef*, report for the Australian Marine Conservation Society, 2021.

35 Mathiesen, Karl, 'Tony Abbott says climate change is "probably doing good"', *The Guardian*, 10 October 2017, theguardian.com/australia-news/2017/oct/10/tony-abbott-says-climate-change-is-probably-doing-good.

36 Brodie, Jon et al., *2013 Scientific Consensus Statement: Land Use Impacts on Great Barrier Reef Water Quality and Ecosystem Condition*, Reef Water Quality Protection Plan Secretariat, 2013, reefplan.qld.gov.au/__data/assets/pdf_file/0018/46170/scientific-consensus-statement-2013.pdf

37 The State of Queensland, *Reef Water Quality Protection Plan 2013: Securing the health and resilience of the Great Barrier Reef World Heritage Area and adjacent catchments*, Reef Water Quality Protection Plan Secretariat, 2013, reefplan.qld.

gov.au/__data/assets/pdf_file/0016/46123/reef-plan-2013.pdf.

38 Zethoven, *The Last Decade*.

39 Ibid.

40 Stockdale, Tim, Magdalena Balmaseda and Laura Ferranti, 'The 2015/2016 El Niño and beyond', European Centre for Medium-Range Weather Forecasts, ecmwf.int/en/newsletter/151/meteorology/2015-2016-el-nino-and-beyond.

Chapter 14: Situation Critical

1 Ridd, Peter, 'Senate inquiry is bringing evidence about state of Great Barrier Reef to the surface', *The Australian*, 16 September 2020, theaustralian.com.au/commentary/senate-inquiry-is-bringing-evidence-about-state-of-great-barrier-reef-to-the-surface/news-story/5e12533a9593e7fd93579cd9b68ee0aa.

2 AIMS, *Long Term Monitoring Program – Summary Report of Coral Condition 2020/2021*, AIMS, 2021.

3 Cornwall, Christopher, et al., 2021, 'Global declines in coral reef calcium carbonate production under ocean acidification and warming', *Proceedings of the National Academy of Sciences*, vol. 118, no. 21, doi.org/10.1073/pnas.2015265118.

4 Hardisty, Paul, et al., 'If we can put a man on the Moon, we can save the Great Barrier Reef', *The Conversation*, 24 April 2020, theconversation.com/if-we-can-put-a-man-on-the-moon-we-can-save-the-great-barrier-reef-121052.

5 Kleypas, Joan, et al., 2020, 'Designing a blueprint for coral reef survival', *Biological Conservation*, vol. 257, doi.org/10.1016/j.biocon.2021.109107.

6 Anthony, Kenneth, et al., 'Interventions to help coral reefs under climate change – A complex decision challenge', *PLOS One*, 26 August 2020, doi.org/10.1371/journal.pone.0236399.

7 Osipova, Elena, et al., *IUCN World Heritage Outlook 3: A Conservation Assessment of All Natural World Heritage Sites*, IUCN, Gland, Switzerland, 2020.

8 McKenna, Michael, 'Coral repair raises hopes for reef as heritage vote looms', *The Australian*, 19 July 2021.

9 McCutcheon, Peter, 'Australia stopped the Great Barrier Reef being listed as 'in danger' – what happens next?', ABC News, 5 August 2021, abc.net.au/news/2021-08-05/unesco-world-heritage-committee-great-barrier-reef-qld/100353360

10 Ridd, Peter, 'It's reef science that's rotten, not the Great Barrier', *The Australian*, 7 December 2020, theaustralian.com.au/exclusives/its-reef-science-thats-rotten-not-the-great-barrier/news-story/1d95f63bc1a2651f5f7c4f9837750513.

11 Readfearn, Graham, 'World Heritage Committee agrees not to place Great Barrier Reef on "in-danger" list', *The Guardian*, 24 July 2021, theguardian.com/environment/2021/jul/23/world-heritage-committee-agrees-not-to-place-great-barrier-reef-on-in-danger-list

12 Ibid.

Chapter 15: Acts of Resistance

1   Morrill, James, *Sketch of a Residence Among the Aboriginals of Northern Queensland for Seventeen Years*, Courier General Printing Office, Brisbane, 1863.

2   Ibid.

3   Breslin, Bruce, *James Morrill: Captive of Empire*, Australian Scholarly Publishing, Melbourne, 2017.

4   Davenport, John, 1997, 'Temperature and the life-history strategies of sea turtles', *Journal of Thermal Biology*, vol. 22, no. 6, doi.org/10.1016/S0306-4565(97)00066-1.

5   Santidrián Tomillo, Pilar, and James Spotila, 2020, 'Temperature-Dependent Sex Determination in Sea Turtles in the Context of Climate Change: Uncovering the Adaptive Significance', *BioEssays*, vol. 42, no. 11, doi.org/10.1002/bies.202000146.

6   Commonwealth of Australia, *Reef Snapshot: Summer 2021–22*, GBRMPA, 2022.

7   Ridd, Peter, 'The reef is strong, so stop the scare campaign', *The Australian*, 5 August 2022, theaustralian.com.au/commentary/the-reef-is-strong-so-stop-the-scare-campaign/news-story/edc8df5f78b30b4920c0f682ed0faf87.

8   AIMS, *Long-term Monitoring Program Annual Summary Report on Coral Reef Condition 2021/22*, AIMS, 2022.

9   Happer, Catherine, and Greg Philo, 2016, 'New Approaches to Understanding the Role of the News Media in the Formation of Public Attitudes and Behaviours on Climate Change', *European Journal of Communication*, vol. 31, no. 2, doi.org/10.1177/0267323115612213.

10 IPCC, *Sixth Assessment Report: The Physical Science Basis*, University of Cambridge Press, 2021.

11 Veron, John (Charlie), *A Reef in Time*, Harvard University Press, 2008.

12 Harrison, Peter, Steve Dalton and Andrew Carroll, 2011, 'Extensive Coral Bleaching on the World's Southernmost Coral Reef at Lord Howe Island, Australia', *Coral Reefs*, vol. 30, no. 775, doi.org/10.1007/s00338-011-0778-7.

13 Harriott, Vicki, Peter Harrison and Simon Banks, 1995, 'The coral communities of

Lord Howe Island', *Marine and Freshwater Research*, vol. 46, no. 2, doi.org/10.1071/MF9950457.

14 Abdul Wahab, Muhammad Azmi, et al., 2023, 'Hierarchical settlement behaviours of coral larvae to common coralline algae', *Scientific Reports*, vol. 13, no. 1, doi.org/10.1038/s41598-023-32676-4.

15 Hardisty, Paul, David Mead and Rob Vertessy, 'Saving the Great Barrier Reef: these recent research breakthroughs give us renewed hope for its survival', *The Conversation*, 24 March 2022, theconversation.com/saving-the-great-barrier-reef-these-recent-research-breakthroughs-give-us-renewed-hope-for-its-survival-178898.

16 Bay, Line, et al., 2023, 'Management approaches to conserve Australia's marine ecosystems under climate change', *Science*, vol. 381, no. 6658, doi.org/10.1126/science.adi3023.

Chapter 16: An Uncertain Future

1   Painter, James, 'Climate change: multi-country media analysis shows scepticism of the basic science is dying out', *The Conversation*, 18 April 2023, theconversation.com/climate-change-multi-country-media-analysis-shows-scepticism-of-the-basic-science-is-dying-out-198303.

2   Wallen, Scout, 'Remember that record $443m of funding for the Great Barrier Reef? This is what happened to it', ABC News, 22 April 2023, abc.net.au/news/2023-04-22/remember-that-record-funding-for-the-great-barrier-reef/102252268.

3   Harvey, Fiona, 'Scientists deliver "final warning" on climate crisis: act now or it's too late', *The Guardian*, 21 March 2023, theguardian.com/environment/2023/mar/20/ipcc-climate-crisis-report-delivers-final-warning-on-15c.

4   IPCC, *Sixth Assessment Report: Synthesis Report*, University of Cambridge Press, 2023.

5   Carter, Bob, 'There IS a problem with global warming ... it stopped in 1998', *The Telegraph*, 9 April 2006, telegraph.co.uk/comment/personal-view/3624242/There-IS-a-problem-with-global-warming...-it-stopped-in-1998.html.

6   Held, Isaac, 2013, 'The cause of the pause', *Nature*, vol. 501, doi.org/10.1038/501318a.

7   Flannery, Tim, *The Weather Makers*, Text, Melbourne, 2005.

8   Dyer, Gwynne, *Climate Wars*, Scribe, Melbourne, 2010.

9   US Department of Defense, *National Security Implications of Climate-Related Risks*

and a Changing Climate, 23 July 2015.

10 Hoegh-Guldberg, Ove, et al., 2023, 'Coral Reefs in Peril in a Record-breaking Year', *Science*, vol. 382, no. 6676, doi.org/10.1126/science.adk4532.

11 Wysession, Michael, 'Global temperatures are off the charts for a reason: 4 factors driving 2023's extreme heat and climate disasters', *The Conversation*, 27 July 2023, theconversation.com/global-temperatures-are-off-the-charts-for-a-reason-4-factors-driving-2023s-extreme-heat-and-climate-disasters-209975.

12 Willcock, Simon, et al., 2023, 'Earlier collapse of Anthropocene ecosystems driven by multiple faster and noisier drivers', *Nature Sustainability*, vol. 6, doi.org/10.1038/s41893-023-01157-x.

13 Carrington, Damian, '2023 smashes record for world's hottest year by huge margin', *The Guardian*, 9 January 2024, theguardian.com/environment/2024/jan/09/2023-record-world-hottest-climate-fossil-fuel.

14 Reid, Kimberley, 'Why are so many climate records breaking all at once?', *The Conversation*, 6 July 2023, theconversation.com/why-are-so-many-climate-records-breaking-all-at-once-209214.

Chapter 17: Taking Action

1 Copernicus Climate Change Service, 'Copernicus: In 2024, the World Experienced the Warmest January on Record', 8 February 2024, climate.copernicus.eu/copernicus-2024-world-experienced-warmest-january-record.

2 Watts, Jonathan, 'February on course to break unprecedented number of heat records', *The Guardian*, 17 February 2024, theguardian.com/environment/2024/feb/17/february-on-course-to-break-unprecedented-number-of-heat-records.

3 NOAA Coral Reef Watch, 'Australia 5 km Regional Virtual Station Time Series Graphs for Lord Howe Island, Australia (v3.1)', March 2024, coralreefwatch.noaa.gov/product/vs/timeseries/australia.php#lord_howe_island.

4 ——, 'Daily Global 5km Satellite Coral Bleaching Heat Stress Alert Area (v3.1)', March 2024, coralreefwatch.noaa.gov/product/5km/index_5km_baa-max-7d.php.

5 Dickie, Gloria, 'Exclusive: World on brink of fourth mass coral reef bleaching event, NOAA says', *Reuters*, 5 March 2024, reuters.com/business/environment/world-brink-fourth-mass-coral-reef-bleaching-event-noaa-says-2024-03-05/.

6 CSIRO, *2019 Australian National Outlook*, CSIRO Publishing, 2019.

7 Hardisty, *Environmental and Economic Sustainability*.

8 Ibid.